UNIVERSITY OF CALIFORNIA
COLLEGE OF CHEMISTRY
BERKELEY, CALIFORNIA 94720

ATLAS OF ELECTRONIC SPECTRA OF
5-NITROFURAN COMPOUNDS

J.A. EIDUS, A.Ya. EKMANE,
K.K. VENTERS and S.A. HILLER

ATLAS OF ELECTRONIC SPECTRA OF 5-NITROFURAN COMPOUNDS

Translated by J. SCHMORAK

ISRAEL PROGRAM FOR SCIENTIFIC TRANSLATIONS

ANN ARBOR SCIENCE PUBLISHERS

ANN ARBOR · LONDON · 1970

CHEM

ANN ARBOR SCIENCE PUBLISHERS, INC.
Drawer No. 1425, 600 S. Wagner Road, Ann Arbor, Michigan 48106

ANN ARBOR SCIENCE PUBLISHERS, LTD.
5 Great Russell Street,   London W.C. 1

Library of Congress Catalog Card Number 70–135515
SBN 250 97504 1

This book is a translation from Russian of
ATLAS ELEKTRONNYKH SPEKTROV
5-NITROFURANOVYKH SOEDINENII
Izdatel'stvo "Zinatne"
Riga, 1968

Preface

The derivatives of 5-nitrofuran are now extensively used in medicine and in veterinary medicine.

This atlas comprises the electronic absorption spectra of fifty 5-nitrofuran derivatives, mainly 2-polyene-substituted compounds. The spectra are plotted both on the linear and on the logarithmic intensity scale. The spectral bandwidth is shown for each absorption band to enable quantitative measurements to be performed.

The descriptive part includes general information on the electronic absorption spectra of 5-nitrofurans, as well as a number of general relationships derived by one of the authors from the experimental material in the atlas.

This book is intended for chemists and physicists specializing in heterocyclic compounds. The experimental work has been carried out at the Institute for Organic Synthesis, Latvian Academy of Sciences, and at the P. Stuchka Latvian State University.

Table of Contents

viii

Introduction

Furan and furan derivatives have been known to chemists for a long time. "Pyromucic acid" (2-furancarboxylic acid) was isolated by Scheele [1] as early as 1780 by dry distillation of mucic acid. Furfural was isolated in 1832 by Döbereiner [2]. Furan was first prepared by Limpricht [3] in 1870, while the first nitro derivatives of furan were prepared by Klinkhardt [4] in 1882.

However, the practical importance of the furan compounds was not realized until much later. 5-Nitrofurans substituted in the 2-position became very important after their biological activity had been discovered in 1944 by Dodd, Stillman, Roys and Crosby [5].

More than 1000 papers on the chemistry of nitrofurans and about 5000 papers on experimental and clinical studies of 5-nitrofuran derivatives in medicine and veterinary medicine have been published during the past 20 years.

In 1946 a systematic search for new preparations of the 5-nitrofuran series and the study of their reactions and properties were launched at the Latvian Academy of Sciences. Experiments showed that the most active nitrofuran compounds are those with conjugated vinylene groupings in the side chain, i.e., 5-nitrofuryl-2-polyenes. Thus, for instance, the original Soviet-made drug Furagin (1-[β-(5'-nitrofuryl-2')-acrylideneamino]-hydantoin) is successfully used for medical purposes and is much more active than the parent compound of the series, Furadonin (1-(5'-nitro-2'-furfurylideneamino)-hydantoin).

5-Nitrofuran polyenes have now become of increasing scientific and practical interest, and accurate experimental data are needed to make possible their analytical identification and quantitative determination.

This is the main reason for the compilation of the present atlas, which will be found useful for spectrophotometric analysis of 5-nitrofurans.*

These compounds are also of interest as derivatives of heteroaromatic five-membered-ring compounds, and we accordingly give a discussion of the general electronic structure of furan and its 5-nitro derivatives.

Some of the material presented in the atlas has been previously published in scientific periodicals or as reports presented to scientific meetings [6–21].

The main subject of this book is the series of 2-polyene-substituted 5-nitrofurans of the general formula

$$O_2N-\underset{O}{\boxed{}}-(CH{=}CH)_n-X,$$

where X denotes the various substituents, while n is an integer between 0 and 4.

Intramolecular coupling effects, both in the immediate vicinity of the functional group and between the parts of the molecule separated by conjugated vinylene groups and the furan ring, are of theoretical interest. We consider the effect on the furan ring of the 5-nitro group and of other electron-accepting substituents in position 2, and also the interaction between the functional groups in 5-nitro-2-furylpolyenes. Other topics discussed include the nature of the electronic absorption bands of 5-nitro-furan compounds and the aromatic nature of the furan ring.

The furan ring itself is the main constituent of the compounds listed in this atlas; we shall accordingly begin by giving a brief description of its structure and properties.

The treatment and interpretation of the spectra in this book is based on the theory of molecular orbitals. We shall take this opportunity to reiterate the fundamental assumptions of this theory which will be used in the following treatment, and to explain the notation employed.

---

* For a detailed survey of methods of spectrophotometric analysis of 5-nitrofuran drug preparations see V. E. Egerts, J. Stradiņš and M. V. Shimanska, *Metody analiticheskogo opredeleniva soedinenii 5-nitrofuranovogo ryada* (*Analytical Determination of 5-Nitrofuran Compounds*) — "Zinatne", Riga, 1968. [English translation by Ann Arbor Science Publ. 1970.]

From the point of view of the chemical bond theory, the relevant electronic states are those corresponding to σ, σ^*, π, π^* and n orbitals, where

σ is the strongly bonding molecular orbital, consisting of combinations of s electrons (s orbital) and z electrons (i.e., p orbital oriented in the direction of the bond). A single bond between two atoms is realized by means of σ orbitals. The electrons here are largely localized. Thus, a relatively high energy is required to excite the electrons;

σ^* is the corresponding anti-bonding orbital;

π is the bonding molecular orbital consisting of a combination of atomic x or y electrons (i.e., a p orbital perpendicular to the bond and to the plane of the molecule (x) or lying in the plane of the molecule (y)). These orbitals are involved in the formation of double bonds; their electrons are localized to a lesser extent than the σ electrons, and less energy is required to excite a π electron than a σ electron;

π^* is the corresponding anti-bonding orbital;

n are the nonbonding orbitals, which participate in bond formation only to an insignificant extent and are localized on one atom (e.g., the lone-pair electrons of the hetero atom in the carbonyl group or in pyridine, which do not participate in conjugation; this is the case, in particular, in furan or aniline). These orbitals have no corresponding anti-bonding states.

In electron transitions, the electron is raised to a state of higher energy. Depending on the molecular orbitals which correspond to the initial and the final states in the transition, we distinguish between several types of electron transitions. Thus, a transition from a bonding orbital in the ground state to a higher energy bonding or anti-bonding orbital is denoted by showing the initial and the final states, e.g., $\pi \rightarrow \pi$ or $\pi \rightarrow \pi^*$, or, in general, as a $N \rightarrow V$ transition (i.e., a transition from the ground (normal) state in the valent shell). If the transition takes place from a nonbonding orbital localized on a given atom to a higher energy orbital, it is also convenient to indicate the initial and the final states of the transition ($n \rightarrow \pi^*$, etc.) or, more generally, $N \rightarrow Q$.

For more information on the subject, the reader is referred to special literature on the theory of electronic spectra of molecules, e.g. [22].

PART I

Electronic Absorption Spectra of 5-Nitrofurans and Certain Relationships for the 5-Nitrofuryl-2-Polyene Series

The Structure of the Furan Molecule

The molecule of furan is a planar, oxygen-containing, five-membered ring. Thompson and Temple [23] established the symmetry of furan (C_{2v}) and demonstrated by the spectroscopic technique that the furan molecule is planar. The furan molecule was repeatedly studied by electron diffraction. The most accurate results, recently obtained by Bak and Christensen [24], yield the following values for the geometric parameters of the molecule:

Bond lengths	Bond angles
$OC_2 = 1.362$ Å	$C_5OC_2 = 106°33'$
$C_2C_3 = 1.361$ Å	$OC_2C_3 = 110°41'$
$C_3C_4 = 1.431$ Å	$C_2C_3C_4 = 106°03'$
$C_2H = 1.075$ Å	$OC_2H = 115°55'$
$C_3H = 1.077$ Å	$C_4C_3H = 127°57'$

The lone pair electrons of the oxygen atom probably also participate in the formation of C—O bonds, which accordingly display certain features of double bonds. The shorter C—C bond length (the "normal" length of the C—C bond in paraffins is 1.54 Å) is probably due to a similar effect.

3

This, in turn, results in the C=C bonds becoming somewhat longer than in olefins, where the "normal" bond length is 1.34 Å.

The four electrons of the C=C double bonds, together with the lone pair electrons of oxygen, form the aromatic sextet. However, since an oxygen atom is more negative than, say, a nitrogen atom, a sulfur atom or a selenium atom, the furan molecule is not as strongly aromatic as the other five-membered-ring heterocyclic compounds — pyrrole, thiophene and selenophene. In fact, electron diffraction data obtained by Schomaker and Pauling [25] for the C—O bond lengths in the furan molecule show that the charge delocalization in the molecule does not exceed 10%, while the corresponding figures are 24% in pyrrole and 34% in thiophene.

The spectroscopic data and the results of calculations indicate that the furan molecule is only insignificantly aromatic.

Furan thus should be regarded in most cases as a cyclic diene ether, with only weak aromatic properties.

However, the aromatic nature of the furan ring must not be totally disregarded. Thus, when furan is hydrogenated [26], the C—O—C bond angle becomes 110° in 2,5-dihydrofuran and 111° in tetrahydrofuran, while the C—C bond length is 1.54 Å in both cases, which is the same as the corresponding bond length in paraffins. The length of the C—O bonds also increases (1.45 Å in 2,5-dihydrofuran and 1.43 Å in tetrahydrofuran) [27].

Electronic Absorption Spectra of Furan

Carefully purified furan absorbs in the far ultraviolet, and only a weak plateau is observed at 260–280 mμ, in the middle ultraviolet (Figure 1). The spectra of furan given in a number of handbooks and monographs, including the tables of Landolt–Börnstein [28] and Gillam and Stern [29, p. 188] etc., are erroneous. These sources indicate a weak band (log ε < 0) with pronounced vibrational structure in the middle ultraviolet. This spectrum was first obtained by Menczel [30] and it was subsequently cited by other authors. Other workers [31, 22] reported continuous absorption of furan in this range. We, too, obtained a spectrum agreeing with that of Menczel for our first samples, but subsequently found that the vibrational

4

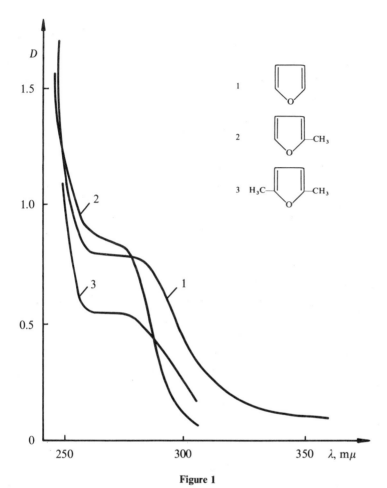

Figure 1

Absorption spectra of furan and some alkylfurans.

band was associated with aromatic impurities present in small amounts [32]. The presence of impurities was confirmed by the appearance of a Raman band at $1000 \, \text{cm}^{-1}$ at low exposure. In the spectrum of pure furan, this band ($994 \, \text{cm}^{-1}$) is weak and does not appear at such exposures.

The problem of the electronic spectrum of furan was recently studied by Horváth and Kiss [33]. In agreement with our own conclusions, these workers attributed the weak band at $260 \, \text{m}\mu$ to aromatic impurities and

5

reported, in perfect agreement with our own results, that only a small plateau is observed at this wavelength in carefully purified furan. This plateau could correspond to a singlet-triplet transition, but such transition was observed only for thiophene at 320 mμ [34]. The solution of this problem is made more difficult by the absence of fluorescence in monoheterocyclic compounds. Kasha [35] discussed a possible radiationless transition from the excited π^* state to the lower metastable triplet level in nitrogen-containing heterocyclic compounds, which could result in phosphorescence.

The intense band with a maximum around 215 mμ has been attributed [33] to a $\pi \rightarrow \pi^*$ transition. The band corresponding to a $n \rightarrow \pi^*$ transition (excitation of the lone electron pair of the hetero atom) does not appear as a discrete formation and seems to merge with the more intense $\pi \rightarrow \pi^*$ transition band.

Pickett [31] studied the spectrum of furan in the far ultraviolet. The structure of the spectrum is distinctly vibrational, and analysis yielded three vibration frequencies of the excited molecule: 1495, 1068, and 848 cm^{-1}. The main electronic absorption bands are at 185, 166, and 140 mμ. The electronic absorption spectrum of cyclopentadiene [36, 37], with absorption bands at 195 and 165 mμ and at a third unspecified wavelength, clearly resembles the spectrum of furan.

The electronic absorption spectrum of furan shows only minor changes as a result of introduction of low-activity substituents. Thus, the absorption spectra of alkylfurans are almost undistinguishable from the spectrum of furan itself [38, 39]. This is confirmed by our own absorption spectra of 2-methylfuran (silvan) and 2,5-dimethylfuran (Figure 1).

2-Nitrofuran

If the highly electron-active nitro group is introduced as a substituent in the furan ring, the absorption spectrum of the resulting compound differs markedly from that of furan. The introduction of such groups produces a major bathochromic effect. A similar effect is also given by 2-vinylfuran [40], where the bathochromic shift is produced by the introduction of an additional conjugated link. The absorption band maximum of 2-vinylfuran is around 260 mμ.

The electronic absorption spectrum of 2-nitrofuran (Figure 2) has two wide bands,† of which the longer wavelength band is the stronger. The two bands are located at 226 and 302 mμ. *The absorption spectrum of 2-nitrofuran has all the typical fingerprints of the absorption spectra of nitrofuran compounds,* both as regards its shape and the relative intensities of the two bands.

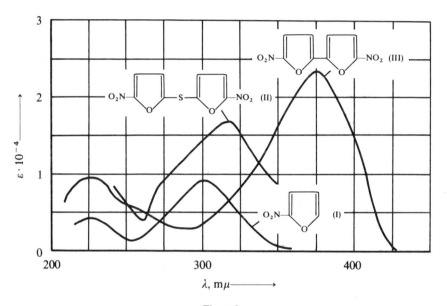

Figure 2

Absorption spectra of 2-nitrofuran (I), 5,5'-dinitro-2,2'-difuryl sulfide (II), and 5,5'-dinitro-2,2'-difuryl (III).

The nitro group is a chromophore with an absorption band around 275 mμ. This relatively weak band (log ε = 1.00–1.18) appears in saturated nitro compounds [41–44]. Nitro compounds of this type also have a strong absorption band at shorter wavelengths, around 210 mμ. According to [45], the weak long-wave band corresponds to a "forbidden" transition $N \rightarrow Q\,(n \rightarrow \pi^*)$, while the strong short-wave band corresponds to an allowed transition $N \rightarrow V\,(\pi \rightarrow \pi^*)$ (Figure 3).

† The electronic absorption spectrum of 2-nitrofuran was first obtained by Raffauf [41]. His results are somewhat different from ours.

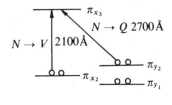

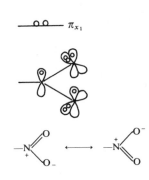

Figure 3

Orbitals and energy levels of the nitro group.

A different situation is created if the nitro group is conjugated with the vinylene group C=C. The grouping $-C=C-N\begin{smallmatrix}\nearrow O \\ \searrow O^-\end{smallmatrix}$ contains a π-electron system in which the lower unfilled π level (π_3) is below the anti-bonding π level (π^*) of the isolated nitro group, and the absorption band is therefore shifted toward longer waves. The corresponding transition may be written as $\pi_1^2\pi_2^2 y_0^2 \rightarrow \pi_1^2\pi_2^2 y_0 \pi_3$.

The band of the $N \rightarrow V$ transition of the nitro group is shifted toward longer wavelengths. The "forbidden" band, corresponding to the $N \rightarrow Q$ transition, shifts toward shorter wavelengths, overlaps the intense K band, moves past it or disappears altogether.

The resulting spectrum is markedly altered, and it may therefore be assumed that an intense band is present at longer wavelengths. The results which were obtained by Braude, Jones and Rose for olefins [46] confirm these ideas.

When the conjugated chain becomes longer, i.e., in structures such as —C=C—C=C—NO$_2$ or Ar—C=C—NO$_2$, the strong band corresponding to the $N \rightarrow V$ transition is further shifted toward the red end of the spectrum. A two-band system is formed, one (the stronger band) around 300 mμ, the other (the weaker band) around 225 mμ [46]. Our own measurements of the spectra of α-nitrobutadiene and β-nitrostyrene ($\lambda_{max} = 300$ and 225 mμ) are a convincing confirmation of this fact.

Thus, the spectrum of dienic nitro compounds is fully analogous to the spectrum of 2-nitrofuran; this is an important finding, to which we shall return below.

In addition to the spectrum of 2-nitrofuran (I), we also took the spectra of 5,5'dinitro-2,2'-difuryl sulfide (II) and 5,5'-dinitro-2,2'-difuryl (III) (see Figure 2). We think that compound (II) is of interest as an example of two nitrofuryl chromophore radicals separated by a "non-conducting" sulfur bridge. As might have been expected, the intramolecular interaction of these chromophores is negligible and the molecule absorbs light at approximately the same wavelength as 2-nitrofuran (with a small bathochromic shift); the absorption intensity is doubled, however. This is in agreement with the literature data on the additivity of the absorption intensities in polynitro compounds [41, 42, 47]. The carbonyl group also obstructs the intramolecular interaction, but to a weaker extent than the sulfur atom. For this reason, the spectrum of 1,5-bis-(5'-nitrofuryl-2')-pentadien-1,3-one-3 displays a slight bathochromic shift with respect to the spectrum of β-(5-nitrofuryl-2)-acrolein (Figures 16 and 62). For further information see [48, 49].

5,5'-Dinitro-2,2'-difuryl presents an altogether different picture. The long-wave band shows a strong red shift and a greatly increased intensity. It is significant that the short-wave band shows practically no shift, but becomes about twice as strong. This marked shift of the spectrum may be explained by the presence of the conjugated "tetraene" chain between the two nitro groups.

These conclusions are in agreement with the recently published results of Wynberg and van Reijendam [50] who prepared and studied 3,3'-difuryl and also studied the previously known 2,2'-difuryl, comparing their spectra with those of the compounds 2,3-divinyl-1,3-butadiene and 1,3,5,7-octatetraene, respectively.

9

These analogies are based on the following bond systems:

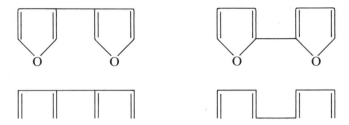

The relevant spectroscopic parameters are as follows:

Compound	λ_{max}, mμ	ε_{max}
3,3'-Difuryl	231	7,780
2,3-Divinyl-1,3-butadiene	223	20,800
2,2'-Difuryl	281	18,500
1,3,5,7-Octatetraene	304	(no available data)

2-Monosubstituted Furan Derivatives with a Carbonyl Group

If a carbonyl-containing group is substituted in the 2 position of the furan ring, the resulting spectrum is expected to be very similar to the spectrum of 2-nitrofuran, owing to the specific properties of the carbonyl group. In fact, the spectra of furfural* and of 2-acetylfuran resemble the spectrum of 2-nitrofuran, except that the bands show a small shift toward shorter wavelengths. One band only is given [42] by the ethyl ester of 2-furan-carboxylic acid and by furfural semicarbazone.

The electronic properties of the carbonyl group greatly resemble those of the nitro group [45]. A single carbonyl group in saturated aldehydes and ketones, like the nitro group, displays a strong absorption band in the far

---

* The spectrum of furfural was first obtained by Dobrinskaya et al. [51, 52].

UV: this band probably corresponds to the $N \to V$ transition $\sigma^2\pi_x^2 y_0^2 \to$
$\to \sigma^2\pi_x^2 y_0 \pi_x^*$, which may overlap the $N \to Q$ transition $\sigma^2\pi_x^2 y_0^2 \to \sigma^2\pi_x^2 y_0 \sigma^*$.
In addition to this allowed transition, an isolated carbonyl group, like
the nitro group, has a "forbidden" transition which gives a weak long-wave
band (around 270 mμ in acetone and formaldehyde). This is the $N \to Q$
transition $\sigma^2\pi_x^2 y_0^2 \to \sigma^2\pi_x^2 y_0 \pi_x^*$ forbidden by the selection rules [22, p. 553]
(Figure 4).

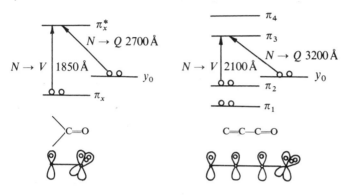

Figure 4

Orbitals and energy levels of a carbonyl group and of a conjugated carbonyl group.

If a vinylidene group is conjugated with the carbonyl group, in α, β-
unsaturated aldehydes, say, the absorption bands display a bathochromic
shift. The effect is intensified if the chain of conjugated C=C groups becomes
longer. Thus, hexadien-2,4-al has a strong absorption band at 270 mμ
[54, 55], which corresponds to the absorption range of furfural [51].

5-Nitro-2-Substituted Furans

If the hydrogen atoms of the furan ring are replaced with a 5-nitro group
and at the same time by a functional group in position 2, new possibilities
arise for the study of the intramolecular interaction *via* the furan ring. It
was shown [56] that this configuration of the substituents is similar to a
certain extent to the *para*-substitution in the benzene ring.

Our study of the substituents in position 2 included radicals with a carbonyl group and radicals with hydrazone grouping.

1. Effect of Carbonyl-Containing Substituents on Electronic Spectra of 5-Nitrofurans

We studied the 5-nitrofuran derivatives with a carbonyl-containing substituent in position 2, listed in Table 1.

In view of the above, it should be expected that the simultaneous introduction of two active bathochromic groups in positions 2 and 5 would necessarily result in a strong bathochromic shift of the spectrum with respect to the spectrum of 2-nitrofuran (Table 1, No. 1), owing to the increase

Table 1

5-NITROFURANS WITH A CARBONYL-CONTAINING SUBSTITUENT
IN POSITION 2 OF THE FURAN RING

No.	Compound, $R = O_2N$—[furan ring]—	$\lambda_{1\ max}$, mμ	$\log \varepsilon_1$	$\lambda_{2\ max}$, mμ	$\log \varepsilon_2$
1	R—H	302	3.95	226	3.56
2	R—CH‖O	310	4.05	226	3.58
3	R—C(‖O)—CH$_3$	299	4.01	221	3.89
4	R—C(‖O)—OH	305–310	4.04	213	4.03
5	R—C(‖O)—OCH$_3$	295	4.12	—	—
6	R—C(‖O)—OC$_2$H$_5$	296	4.05	213	3.88
7	R—CH(OCCH$_3$)$_2$‖O	298	4.06	226	3.72

in the length of the conjugated chain. Table 1 shows, however, that this is not the case and that the absorption spectra of all the compounds are the same as the spectrum of 2-nitrofuran.

Carbonyl-substituted compounds (Table 1, Nos. 3–7) display a slight hypsochromic shift with respect to 5-nitrofurfural (Table 1, No. 2). This effect is typical of carbonyl-containing compounds and is discussed in the literature. Thus, according to Mattsen [22], an exchange of the aldehyde hydrogen for an alkyl or an alkoxy group should shift the spectrum toward shorter wavelengths. Both resonance and inductive effects are operative here. The resonance effect increases the energy of the π_x^* level (Figure 4).

The original electron orbital y_0 is not markedly changed and the spectrum should therefore show a hypsochromic shift. On the other hand, there also appears a positive inductive effect which increases the energy of y_0, which in turn should lower the ionization energy and cause a red shift of the spectrum, i.e., the resonance effect is partly compensated. As a result, the hypsochromic shift of the spectrum should be small.

However, the spectrum of 5-nitrofurfural, too, is not much shifted with respect to the spectrum of 2-nitrofuran. The explanation may be that both groups — the carbonyl and the nitro group — are electron acceptors. Thus, they compete with each other to some extent, and the conjugation is weakened. This offsets the lengthening of the chain and the spectrum of 5-nitrofurfural is only slightly shifted with respect to that of 2-nitrofuran:

It is significant that when an electron-donating group appears in position 2 of the furan ring, a bathochromic shift with respect to the spectrum of 2-nitrofuran is noted. Thus, the strong absorption band of 5-nitro-2-methyl-furan has a maximum at 309 mμ (log ε = 4.04), i.e., it displays a 7 mμ shift with respect to 2-nitrofuran. A similar effect is noted in the case of 5-nitro-2-bromofuran (λ_{max} = 330 mμ, log ε = 3.78).

It is interesting to compare the absorption data for 5-nitro-2-substituted furans with the corresponding parameters for furans and unsaturated aliphatic compounds (see table on p. 14).

Compound, R = O₂N–furan–, R = furan–	λ_{max}, mμ	log ε
R—COOH	305	4.04
R—COOCH₃	295	4.12
R—COOC₂H₅	296	4.05
R′—CH=CH—COOH*	300	4.70
CH₃—CH=CH—CH=CH—CH=CH—COOH**	294	4.56

* The spectrum was also obtained by Hausser et al. [57].
** See [55].

Similar data were obtained for aldehydes [55].

The analogy between pure aliphatic polyenes and polyenes of the furan series has been discussed in [54] and it was concluded that if aliphatic polyenes are to be compared with furyl polyenes, the unsaturated bonds in the carbonyl group and in the furan ring must be added to the number of ethylenic bonds in the open chain (n). Thus, the overall value of n_f for the vinyl analog of furylcarboxylic compounds is given by the expression

$$n_f = n + 3.$$

The presence of a 5-nitro group in the furan ring is equivalent to another ethylenic bond, so that

$$n_{nf} = n_f + 1 = n + 4.$$

2. Effect of Substituents with Hydrazone Group on the Electronic Spectra of 5-Nitrofurans

It is known that 5-nitrofuran derivatives with the hydrazone grouping —CH=NN⟨ in position 2 are particularly valuable therapeutic drugs. Spectroscopic studies were carried out on these important compounds in order to be able to apply the results to analytical purposes.

Compounds in this group are strongly conjugated systems, in which the conjugation is realized both *via* the π electrons of the C=C bonds and *via* the lone electron pairs on nitrogen atoms. The nature of the substituents suggests strong electronic interaction and a pronounced shift of the electron cloud.

The lengthening of the conjugated chain and the strong inductive effect exerted by the electrophilic nitro group, which takes place in accordance with the scheme below (drawn for 5-nitro-2-furfurylidene semicarbazone), should result in a bathochromic shift of the absorption spectrum, as is in fact the case:

The compounds which have been studied and the spectroscopic data are listed in Table 2.

It is seen from the table that the absorption spectra of 5-nitrofurans containing the grouping —CH=N—N$\diagup$$\diagdown$ display a strong bathochromic shift with respect to the carbonyl derivatives. The long-wave absorption band is shifted by 50–60 mμ, which is almost equivalent to the effect of two additional vinylene groups in the side chain of the molecule (see Sec. 1 above).

It has been shown that the main electronic absorption band of the carbonyl derivatives of 5-nitrofurans is quite stable and lies between 295 and 310 mμ in all the compounds studied. Similarly, the main electronic absorption band of the compounds discussed in the present section is at 354–367 mμ (Table 2, Nos. 1–6) or else shows a small additional bathochromic shift (Table 2, Nos. 7–9).

If the atom bound by a double bond to the carbon atom of the semicarbazide group is exchanged for another, the absorption band is shifted (cf. Table 2, Nos. 1, 7 and 9). Thus

$X = O, \quad \lambda_{max} = 367$ mμ;
$X = NH, \lambda_{max} = 385$ mμ;
$X = S, \quad \lambda_{max} = 381$ mμ.

15

Table 2

HYDRAZINE DERIVATIVES OF 5-NITROFURFURAL

No.	Compound, R = O_2N—(furyl)—	$\lambda_{1\ max}$, mμ	$\log \varepsilon_1$	$\lambda_{2\ max}$, mμ	$\log \varepsilon_2$	$\lambda_{3\ max}$, mμ	$\log \varepsilon_3$
1	R—CH=NNH—CO—NH$_2$	367	4.28	264	4.10	—	—
2	R—CH=NNH—CO—(pyridyl)N	360	4.33	277	3.99	217	4.04
3	R—CH=NNH—CO—CH$_2$CN	354	4.21	244	4.03	—	—
4	R—CH=NNH—CO—CH$_2$—NH$_2$	360	4.04	—	—	—	—
5	R—CH=N—N—CH$_2$ (O=C C=O, NH)	366	4.29	273	4.05	—	—
6	R—CH=N—N—CH$_2$ (O=C CH$_2$, O)	356	4.32	262	4.12	—	—
7	R—CH=NNH—CS—NH$_2$	381 (wide maximum)	4.34	289	4.15	239	3.99
8	R—CH=NNH—CO—CHCl$_2$	375	3.95	250	3.80	—	—
9	R—CH=NNH—C—NH$_2$·$\tfrac{1}{2}$H$_2$SO$_4$ (NH)	385	4.28	264	4.10	—	—

This effect may be due to the electronegativity of the atom X, which alters the state of the π-electron cloud throughout the molecule. In fact, the bathochromic shift decreases with increasing electronegativity of X, and vice versa. The weakening of the attractive forces of the carbonyl oxygen (a relatively strong electron acceptor) as a result of its replacement by sulfur or nitrogen produces a more pronounced shift of the entire π-electron system towards the nitro group. This is the cause of the bathochromic shift in the spectrum.

Spectroscopic shifts resulting from changes in the end groups of the substituent can also be explained in this way even if the end group is separated from the other parts of the molecule by a carbonyl group which is known to be a poor transmitter of conjugation (see, for example, Table 2, Nos. 1 and 3).

Aromatic Nature of the Furan Ring

It has already been stated that studies of electronic absorption spectra of 2-monosubstituted and 2,5-disubstituted furans reveal an analogy between these furan derivatives and the corresponding conjugated aliphatic systems of the polyene type. Thus, there is full similarity between the absorption spectra of 2-nitrofuran, α-nitrobutadiene, and β-nitrostyrene (in the last case, the phenyl end group is equivalent to one ethylene bond), on the one hand, and the spectra of other dienic compounds, on the other. In almost all cases the —C≡C—C≡C— bond system of the furan ring proves equivalent to conjugated ethylenic bonds.

Pappalardo [58] made comparative studies of electronic spectra of various five-membered heterocyclic compounds and found that the aromatic character of the furan ring was only weakly pronounced.

These observations are in good agreement with the fact that conjugation through a furan ring enhances the Raman lines to a much greater extent than conjugation through a benzene ring. This was established by Bobovich and Perekalin [59] and also by one of us [13, 14].

Thus, spectroscopic data indicate that the furan ring is more of an aliphatic diene than an aromatic compound. Furan should be regarded essentially as a dienic cyclic ether, in which the intramolecular effects are transmitted through a system of conjugated ethylenic bonds.

The aliphatic diene character of the furan ring is confirmed by numerous data on the reactivity of furan compounds, for example, the 2,5-addition of alkoxy and acyloxy groups and of acetyl nitrate to the ring [26, 60, 61], and also by other reactions typical of dienes.

We have already mentioned that Pauling and other workers, who studied the bond lengths in the furan ring, arrived at similar conclusions.

However, the aromatic properties of the furan ring are not altogether

17

ignorable. Our own studies of the vibrational spectra of furan compounds (constancy of ring vibration frequencies in 2- and 5-substituted derivatives) [14] and chemical data show that furan is to some extent aromatic. This is also indicated by the planar structure of the furan molecule and by the presence of the π-electron sextet. The C—C bonds in the furan ring are shorter than in paraffins and the C=C bonds longer than in olefins. Calculations by the MO LCAO method show that the charge delocalization on the oxygen atom is about 10%.

In view of all the above, it must be concluded, however, that the furan ring is predominantly aliphatic, and not aromatic.

Electronic Spectra
of 5-Nitrofuryl-2-Polyenes

Most of the compounds in this atlas have the general formula

$$O_2N \text{—} \boxed{} \text{—} (CH = CH)_n \text{—} X,$$

where X is the substituent and $n = 0, 1, 2, 3, 4$.

If the conjugated chain is gradually extended, it is possible to study the various intramolecular effects as a function of the conjugated chain length.

5-Nitrofuryl-2-polyenealdehydes (X = CHO) are the only group of compounds within this class which have been spectroscopically investigated. Electronic absorption spectra of these compounds were taken by Saikachi and Ogawa [62]; their results are in agreement with those quoted below.

In the functional derivatives of 5-nitrofuryl-2-polyenes studied by the present authors, the substituents X are mostly the radicals listed in Table 3.

The compounds dealt with in the present chapter are typical asymmetric polyenes, and they are of considerable interest to both physicists and chemists on account of their specific properties. Polyenes are one class of compounds to which quantum-mechanical calculation methods, based on various approximate models, have been successfully applied. The study

Table 3

ABSORPTION MAXIMA IN THE ELECTRONIC SPECTRA OF THE FUNCTIONAL DERIVATIVES OF 5-NITROFURYL-2-POLYENES

Compound $R-(CH=CH)_n-X$, where R is 5-nitrofuryl-2	$n =$	$\lambda_{1\,max}$, mμ	$\varepsilon_1 \cdot 10^{-4}$	$\log \varepsilon_1$	$\lambda_{2\,max}$, mμ	$\varepsilon_2 \cdot 10^{-4}$	$\log \varepsilon_2$	$\lambda_{3\,max}$, mμ	$\varepsilon_3 \cdot 10^{-4}$	$\log \varepsilon_3$
—H (I)	0	302	0.90	3.95	226	0.36	3.56	—	—	—
—CHO (II)	0	310	1.13	4.05	226	0.38	3.58	—	—	—
	1	345	1.92	4.28	240	1.08	4.03	—	—	—
	2	377	3.19	4.50	270	1.61	4.21	225	0.57	3.76
	3	399	4.14	4.62	304	2.39	4.38	234	0.90	3.95
—COCH₃ (III)	0	299	1.02	4.01	221	0.78	3.89	—	—	—
	1	340	2.01	4.30	239	1.38	4.14	—	—	—
	2	377	2.63	4.42	274	1.48	4.17	—	—	—
	3	403	3.57	4.55	305	2.06	4.31	236	0.86	3.93
		425–435	3.68	4.57	330	2.07	4.32	244	0.82	3.91
—COOH (IV)	0	305–310	1.10	4.04	213	1.07	4.03	—	—	—
	1	348	1.35	4.12	234	1.23	4.09	—	—	—
—COOCH₃ (V)	0	295	1.32	4.12	—	—	—	—	—	—
—COOC₂H₅ (VI)	0	296	1.13	4.05	213	0.76	3.88	—	—	—
	1	337–346	1.86	4.27	236	1.31	4.12	—	—	—

19

Table 3 (continued)

Compound $R-(CH=CH)_n-X$, where R is 5-nitrofuryl-2	$n =$	$\lambda_{1\,max}$, $m\mu$	$\varepsilon_1 \cdot 10^{-4}$	$\log \varepsilon_1$	$\lambda_{2\,max}$, $m\mu$	$\varepsilon_2 \cdot 10^{-4}$	$\log \varepsilon_2$	$\lambda_{3\,max}$, $m\mu$	$\varepsilon_3 \cdot 10^{-4}$	$\log \varepsilon_3$
$-COOC_2H_5$ (VI)	2	375–377	2.49	4.40	267	1.56	4.19	—	—	—
	3	390–395	3.41	4.53	294–304	2.27	4.36	230	0.58	3.76
	4	421–424	4.19	4.62	⎰ 319	2.64	4.42	243.	0.82	3.91
					⎱ 324	2.69	4.43			
$-CH(OCOCH_3)_2$ (VII)	0	298	1.15	4.06	226	0.53	3.72	—	—	—
	1	342	1.58	4.20	227	1.75	4.24	—	—	—
	2	378	2.24	4.35	245	1.95	4.29	—	—	—
	3	400	2.84	4.45	286	2.59	4.41	222	0.89	3.95
$-CH=NNHCONH_2$ (VIII)	0	367	1.91	4.28	264	1.27	4.10	—	—	—
	1	389	2.31	4.36	292	1.99	4.30	245	0.86	3.93
	2	413	2.95	4.47	313	2.82	4.45	240	0.79	3.90
	3	425	3.76	4.58	333	3.18	4.50	256	0.79	3.90
$-CH=NNHCSNH_2$ (IX)	0	370–392	2.21	4.34	289	1.41	4.15	239	0.98	3.99
	1	399	2.73	4.44	309	1.90	4.28	254	0.90	3.95
	2	414	2.82	4.45	331	2.12	4.33	232	0.74	3.87
	3	⎰ 421	3.82	4.58	352	2.77	4.44	—	—	—
		⎱ 429	3.77	4.58						
$-(CH=NNHCOCH_2CN$ (X)	0	354	1.62	4.21	244	1.08	4.03	—	—	—

Table 3 (continued)

Compound R—(CH=CH)$_n$—X, where R is 5-nitrofuryl-2 X =		n =	$\lambda_{1\,max}$, mμ	$\varepsilon_1 \cdot 10^{-4}$	log ε_1	$\lambda_{2\,max}$, mμ	$\varepsilon_2 \cdot 10^{-4}$	log ε_2	$\lambda_{3\,max}$, mμ	$\varepsilon_3 \cdot 10^{-4}$	log ε_3
—CH=NNHCOCH$_2$CN	(X)	1	379	2.30	4.36	286	1.93	4.29	246	1.05	4.02
		2	397	2.77	4.44	310	2.48	4.39	240	1.03	4.01
		3	420	4.37	4.64	325	3.14	4.50	240	1.18	4.07
—CH=NNHCO—⟨N⟩	(XI)	0	360	2.15	4.33	277	0.97	3.99	217	1.09	4.04
		1	385	2.73	4.44	301	1.55	4.19	241	1.14	4.06
		2	406	3.82	4.58	320	2.20	4.34	215	1.16	4.06
		3	421	4.28	4.63	344	2.49	4.40	239	0.94	3.97
—CH=N-N-CH$_2$ ring (C=O, C=O, NH)	(XII)	0	366	1.95	4.29	273	1.11	4.05	—	—	—
		1	386	2.44	4.39	291	2.04	4.31	—	—	—
		2	406	2.58	4.41	312	2.15	4.33	242	0.77	3.89
		3	417	2.51	4.40	327	2.07	4.32	—	—	—
—CH=N-N-CH$_2$ ring (C=O, C=O, O)	(XIII)	0	356	2.10	4.32	262	1.31	4.12	—	—	—
		1	380	2.38	4.38	287	1.96	4.29	—	—	—
		2	406	2.90	4.46	309	2.43	4.39	239	0.87	3.94
		3	421	3.27	4.51	325	2.57	4.41	—	—	—
—CH=CH—CO—R	(XIV)		366	3.07	4.49	317	2.14	4.33	—	—	—
—CH=CH—CO—CH=CH—R	(XV)		370	2.98	4.47	239	1.53	4.18	—	—	—

of polyenes may prove very fruitful in the development of the general theory of the chemical bond, and of the general color theory of organic compounds.

Hausser et al. [54, 57, 63–66] were among the first students of the spectroscopy of polyenes. Their classical work, which was performed on polyenealdehydes, polyenecarboxylic acids, diphenylpolyenes, and furyl-2-polyenecarboxylic acids, revealed the general relationships between the color, the absorption, and the structure of the polyene molecule.

The data obtained by Hausser, Kuhn, Smakula and others served as the starting point for the first theoretical interpretation of the observed effects. A large proportion of these studies is based on models which agree with the experimental results to a greater or lesser extent.

One of the first studies of this kind was the investigation of polyene systems carried out by Lewis and Calvin [67]. According to these authors, the π electrons, which have a certain effective charge and a certain effective mass, are displaced by an external electric field, while being subjected to an elastic restoring force. In other words, if a periodic external electric field is applied, the electrons in each π bond will execute harmonic vibrations. If the entire π bond system is quasi-linear (the zigzag-shaped polyene chain may be roughly considered as such), it may be treated as a linear harmonic oscillator of mass nm, where m is the effective mass of the vibrating charge in one bond and n is the number of the vibrating π bonds.

If the oscillator force constant is k', we have for the first vibrational transition of such a system $2\pi v_0 = \sqrt{k'/nm}$, where v_0 is the natural frequency of the characteristic vibrations of the system. Hence

$$\lambda^2 = \frac{4\pi^2 cm}{k'} n = kn ,$$

i.e., the square of the wavelength of the absorption band is proportional to the number of double C=C bonds.

This relationship has been experimentally observed among the symmetrical diphenylpolyenes studied by Hausser et al.

Saikachi and Ogawa [62] found that the relationship is valid for 5-nitro-furyl-2-polyenealdehydes as well. It will be shown below that it also applies to other 2-substituted 5-nitrofurans.

Modern quantum-mechanical calculations, which are mostly based on different variations of the method of molecular orbitals, provide a more accurate description of the true spectroscopic relationships.

Empirical Relationships
in the Electronic Absorption Spectra
of 5-Nitrofuryl-2-Polyenes

The absorption spectra of the 5-nitrofuryl-2-polyenes studied by the authors are assembled in this atlas. The wavelengths of the absorption maxima and the values of the molar extinction coefficients are shown in Table 3.

As was to be expected, all compounds show a regular shift of the main absorption maxima toward longer wavelengths as the number of conjugated double bonds increases. This applies not only to the maxima of the two main absorption bands, but also to the maximum of the third band (corresponding to the shortest wavelength), assuming that it is observable despite the large bathochromic shift of the entire spectrum. The third band appears to be typical of all 5-nitrofuryl-2-polyenes; it is located in the far ultraviolet for molecules with a low degree of conjugation. Quantitatively, the square of the wavelength of the absorption maximum is a linear function of the number of double bonds in the side chain.

This linear relationship, discovered by Saikachi and Ogawa [62] for 5-nitrofuryl-2-polyenealdehydes (in which it is not very accurately obeyed), is consistent with Lewis and Calvin's formula [67]. It is noteworthy that this relationship, which is based on a very primitive model and has been verified for highly symmetric systems such as diphenylpolyenes, also proved valid for asymmetric molecules, in particular 5-nitrofuryl-2-polyenes.

Figure 5 plots the square of the wavelength of the absorption maxima as a function of the number of double bonds in the side chain for ten series of 5-nitrofuryl-2-polyene compounds with 0 to 4 double $C\!\!=\!\!C$ bonds in the side chain.

The corresponding relation may be empirically represented as

$$\lambda_{\max}^2 = a_{ij} + k_{ij}n, \tag{1}$$

where $\lambda_{\max}$ is the wavelength of the absorption maximum of the band, n is the number of conjugated vinylene groups in the side chain, and a_{ij} and k_{ij} are constants characteristic of each absorption maximum and each individual substituent X.

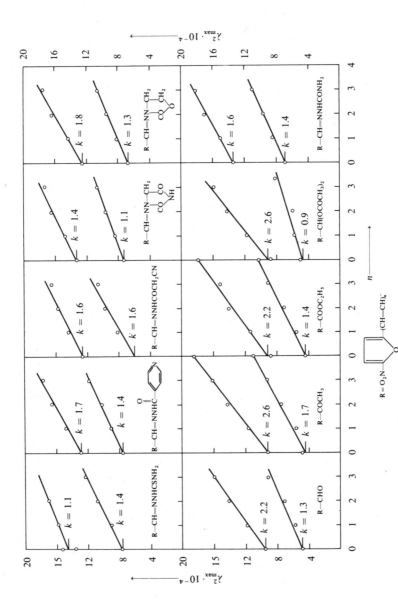

Figure 5

Square of the wavelength of the absorption band maximum as a function of the number of conjugated vinylene groups in the side chain of 5-nitrofuryl-2-polyenes.

The slopes of the straight lines plotted for each absorption band of a given compound — the bathochromic shift of each band for each additional vinylene group — are different. We must accordingly inquire into the nature of the main absorption bands in the electronic spectrum of 5-nitro-furyl-2-polyenes. The different band shifts might be due to the fact that each band is produced by a transition which takes place in a separate electron system ("chromophore").

One such system could be, for example, the 5-nitrofuryl radical, while the other could be $—(CH{=}CH)_n—X$; alternative system pairs would be 5-nitrofuryl-2-polyenyl and a substituent X or else the nitro group and furyl-2-polyenyl-X. In [9] we tried to unravel the electronic absorption spectra of 5-nitrofuryl-2-polyenes assuming the existence of two separate chromophore groups, so that each band was assigned to a certain chromophore group.

However, subsequent investigations have cast serious doubts on the validity of this approach. In the first place, separate chromophores are more usually observed in the presence of "insulating" links in the molecule, whereas all our compounds, which were quite strongly conjugated along the entire molecule, probably had a single π-electron cloud. This is confirmed by Raman and IR absorption spectra [11, 13, 14] and also by theory.

It would appear, accordingly, that the main absorption bands must be regarded as different (say, the first and the second) electron transitions in the same electron system. This point of view is supported by the observed fact that the short-wave band is always weaker than the long-wave band. If this point of view is correct, the different frequency variations of the absorption bands with increasing number of vinylene groups in the side chain should be interpreted as a rearrangement of the system of electronic levels caused by the lengthening of the conjugated chain.

Another strong argument in favor of the interpretation of both absorption bands as the result of transitions within the same π-electron system is provided by the spectrum of 2-nitrofuran itself, in which both the substituent X and the vinylene groups in the side chain are absent. The spectrum shows two absorption bands in the middle ultraviolet (Figure 2), with maxima at 302 and 226 mμ; the respective molar extinction coefficients are $\log \varepsilon_1 = 3.95$ and $\log \varepsilon_2 = 3.56$. This spectrum does not differ qualitatively from the spectra of 2-substituted 5-nitrofurans, which is another argument against the theory of two separate chromophores.

25

The variation of the square of the wavelength of the band maximum with the number of vinylene groups in the side chain is characterized by the constant k_{ij} in equation (1). The value of this constant is an index of the sensitivity of the given electronic transition to the extension of the conjugated chain the the molecule: the longer the chain, the larger is the bathochromic shift and the steeper is the slope of the straight line in Figure 5.

The constant a_{ij} in (1) describes the electronic transition in the molecule of the parent compound, without any vinylene groups in the side chain.

It has been observed that the higher the a_{ij} of the main (long-wave) absorption band, the lower its k_{ij}. This means that the longer the wavelength of the main absorption band, the smaller is the bathochromic shift produced by the introduction of vinylene side groups.

This effect is illustrated in Figure 6, which plots k against a for the long-wave bands of each polyene series with the same substituent X. To a rough approximation, the function is linear.

If vinylene groups are introduced into a molecule with a highly conjugated substituent X, the bathochromic shift will probably be even smaller.

It may be expected that for some highly conjugated substituent X, the spectrum will become "saturated"; the value of the constant will go to zero, and no bathochromic shift will result from the introduction of vinylene groups. This conclusion could not be experimentally confirmed, since sufficiently highly conjugated 2-substituted 5-nitrofurans are not available.

An inspection of Figure 6 makes it possible to estimate the λ_{max} of the parent compound at which the position of the absorption band becomes independent of the length of the conjugated chain of the substituent. In the case of 5-nitrofuryl-2-polyenes, the straight line $k = f(a)$ meets the abscissa at $a \approx 20.5 \times 10^4 \ m\mu^2$, i.e., $k \approx 0$ at $\lambda_{max} = \sqrt{a} \approx 450 \ m\mu$.

If we assume that each ethylene bond or its equivalent produces approximately a 20–30 $m\mu$ shift in the absorption maximum of the parent compound and that for 2-nitrofuran $\lambda_{max} = 300 \ m\mu$, we may expect "saturation" (i.e., $k \approx 0$) for a substituent X equivalent to 5–7 conjugated ethylene bonds. If, in such a compound, vinylene groups are introduced between the 5-nitrofuryl radical and the substituent X, the long-wave band should no longer show a bathochromic shift.

Figure 6 gives the function $k = f(a)$; in our case,

$$k = 4.2 - 0.20a.$$

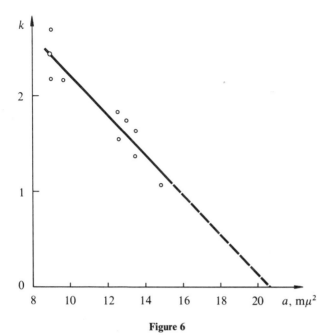

Figure 6

Constant k as a function of the constant a in the spectra of 5-nitrofurans.

If the λ_{max} of the parent compound is known, the formula gives the approximate position of the absorption band for its derivatives with any number of conjugated double bonds in the side chain.

This decrease in the bathochromic effect of vinylene groups introduced into the molecule may be related to a certain extent to Kuhn's "convergence" effect [68] in the spectra of the higher polyenes. Indeed, if a substituent X of a sufficiently complex structure is introduced into the molecule, the resulting additional conjugation is equivalent to a high polyene.

The intensity of the absorption bands increases with increasing number of vinylene groups in the side chain. This effect was theoretically interpreted by various authors. Thus, Kuhn et al. [69] showed that the extinction coefficients of polyenes should increase in proportion to the number of conjugated $C=C$ bonds in the polyene. On the whole, this conclusion was confirmed by our own results (Figure 7).

Platt [70] studied the changes which occur in the intensities of the absorption bands of polyenes as a result of the spectral shift. He found that

27

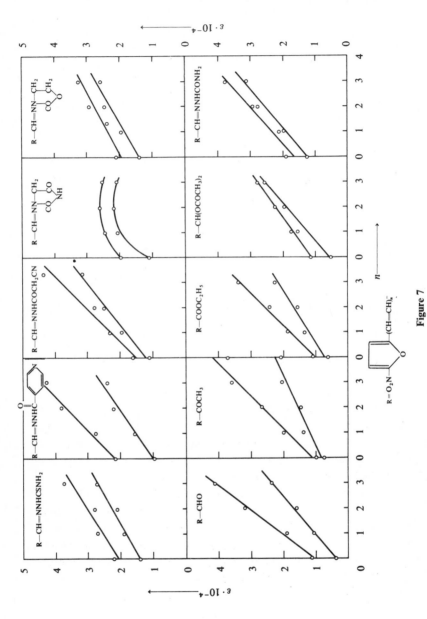

Figure 7

The extinction coefficient ε as a function of the number of double bonds n in the side chain.

28

the intensity increases if the shift is directed toward the red end of the spectrum. The increase is a function of λ_{max}: it is large at first, and then decreases past a certain limit.

In 5-nitrofuryl-2-polyenes, ε increases fairly rapidly with increasing λ_{max} (Figure 8).

However, no decrease in intensity was noted in the series of compounds that we studied. This was probably because the polyene molecules in our work were not long enough.

Some Quantum-Chemical Calculations

We have mentioned that polyenes are a class of compounds particularly suited to various quantum-chemical calculations based on approximate models. Accordingly, the application of these methods to 5-nitrofuryl-2-polyene systems, which form the subject of this book, may be of interest.

1. The Method of the Free-Electron Molecular Orbitals ("The Metallic Model")*

This model, proposed almost simultaneously and independently by Bayliss [71], Simpson [72] and Kuhn [53], proved to be particularly successful when applied to polyenic systems. These authors proceeded from Sommerfeld's theory of the metallic state [73], in which the electrons are considered as a one-dimensional gas in a homogeneous potential field. These simplified assumptions were adopted by Schmidt [74] in his study of the wave functions of electrons in aromatic hydrocarbons, which provided an interpretation of a large number of physiochemical properties of these compounds.

It is assumed that the π electrons moving along the conjugated chain are locked in a potential well ("box model"). The potential in the well varies periodically owing to the presence of the carbon atoms at the chain sites. The nature of this periodicity will be understood on considering the difference between the C—C and C=C bond lengths. If the polyene molecule

---

* In collaboration with P. E. Kunin, I. M. Taksar and L. A. Stolyarova.

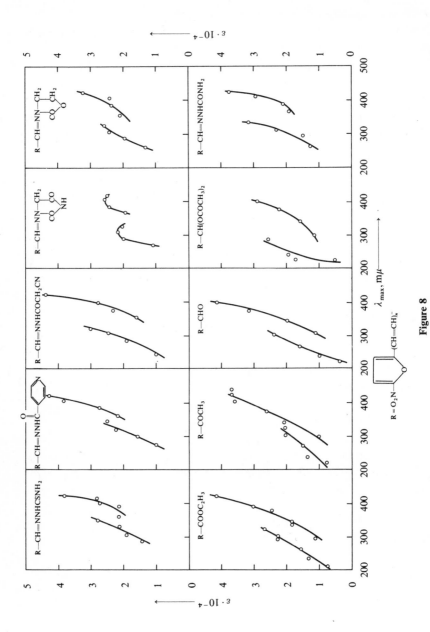

Figure 8

ε as a function of λ_{max}.

30

is symmetric, it may be regarded as two equivalent saturated units, and the periodic potential may be averaged to give a certain constant value. If we assume that the potential at the ends of the molecule rapidly increases to infinity, we have the case of a rectangular well, whose width is determined by the length of the polyene chain. This length can be taken equal to the length of the conjugated chain plus the length of one bond for each additional carbon atom at the end of the chain. The length element is taken as $l = 1.39$ Å, since in the case of equivalent structures all carbon–carbon bonds may be considered as half-double bonds.

Each pair of π electrons in such a potential well occupies one energy level. The first excited state is obtained when an electron jumps from a higher occupied level to a lower unfilled level. The wavelength corresponding to this transition is calculated from the formula

$$\lambda_1 = \frac{8mc}{h} \frac{L^2}{N + 1} \tag{1}$$

where m is the mass of the electron, c is the velocity of light, h is Planck's constant, L is the length of the polyene chain (the width of the potential well), and N is the number of π electrons.

Kuhn introduced certain corrections into (1) in order to allow for the effect of the molecular end groups, but these do not affect the general nature of the relationship between the wavelength of the absorption maximum and the number of π electrons in the conjugated chain.

In the case of asymmetric polyenes, when the saturated structures are not equivalent, it is necessary to allow for the difference in the C—C and C=C bond lengths. The potential distribution at the bottom of the well is then more complex; according to Kuhn, it may be averaged as a sinusoidal curve with an amplitude V_0. In this potential the energy levels are grouped in bands, and the distances between the bands are much larger than the level spacing within any one band. In the ground state, the lower band is filled and the first electron transition therefore occurs between two bands, causing absorption at shorter wavelengths than for a symmetric polyene.

According to Kuhn, the wavelength of the first electron transition is

$$\lambda_1 = \left[\frac{V_0}{hc} \left(1 - \frac{1}{N} \right) + \frac{h}{8mcL^2} (N + 1) \right]^{-1}, \tag{2}$$

where V_0 is the auxiliary potential (the other notation as in (1)). The value of V_0 must be found from the experimental data.

Kuhn's model is used in this atlas for the calculation of spectra of a number of 5-nitrofuryl-2-polyene series which have been studied experimentally. Let us consider the series with the general formula

$$O_2N-\underset{O}{\underset{\|}{\overset{}{\boxed{}}}}-(CH=CH)_n-\underset{O}{\underset{\|}{C}}-CH_3 .$$

If we assume that the two $C=C$ bonds of the furan ring participate in the conjugation, the number of π electrons in the molecule is $N = 2n + 4$, and the chain length (the "width" of the well) is $L = (N + 1)l$ (counting the additional carbon atom in the side chain). For the sake of simplicity l is taken as 1.39 Å, following Kuhn. The calculations are conducted using equation (1) for $V_0 = 0$ and equation (2).

The results obtained for the various values of V_0 are shown in Table 4; the curves plotted from these data and the experimental curve are shown in Figure 9.

Table 4

MAXIMA OF THE LONG-WAVE ABSORPTION BANDS
OF 5-NITROFURYL-2-POLYENEKETONES

n	N	Calculated λ_{max}, $m\mu$							Experimental λ_{max}, $m\mu$
		$V_0 = 0.0$	1.0	1.5	2.0	2.2	2.5	3.0 (eV)	
0	4	318	267	247	230	224	215	202	299
1	6	446	348	308	279	269	255	235	340
2	8	573	409	358	318	304	286	260	377
3	10	700	464	398	347	332	308	277	403
4	12	827	515	433	374	354	328	293	430

Similar data may also be obtained for other series of 5-nitrofuryl-2-polyenes. Thus, we calculated the maximum absorption wavelengths of the

32

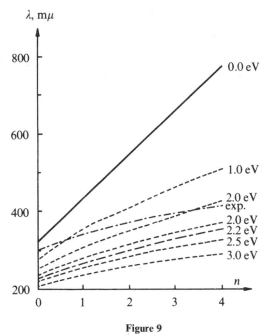

Figure 9

Experimental and theoretical curves of λ_{max} of 5-nitrofuryl-2-polyeneketones as a function of the number of conjugated C=C bonds in the side chain.

5-nitrofuryl-2-polyenealdehyde series of the general formula

$$O_2N-\underset{O}{\bigcirc}-(CH=CH)_n-\underset{\underset{O}{\|}}{C}-H$$

Here, the number of the counted π electrons is also $N = 2n + 4$, and the chain length (the width of the well) is $L = Nl$. The results are shown in Table 5 and Figure 10.

It is seen that equation (1) ($V_0 = 0$), which is valid for a symmetrical molecule, is quite inadequate. Equation (2) gives a set of curves for various values of V_0, none of which coincides with the experimental curve. However, it is possible to find a value of V_0 at which the calculated and the experimental curves are parallel. In the case of ketones (Figure 9), this is $V_0 = 2.2$ eV, and in the case of aldehydes $V_0 = 3.5$ eV. Thus, equation (2)

33

Table 5

MAXIMA OF THE LONG-WAVE ABSORPTION BANDS
OF 5-NITROFURYL-2-POLYENEALDEHYDES

n	N	Calculated λ_{max}								Experimental $\lambda_{max}, m\mu$
		$V_0 = 0.0$	1.0	1.5	2.0	2.5	3.0	3.5	4.0 (eV)	
0	4	204	182	172	163	156	149	143	132	310
1	6	327	268	246	228	212	198	185	174	345
2	8	453	344	307	276	253	232	215	199	377
3	10	579	408	355	315	282	256	234	216	399

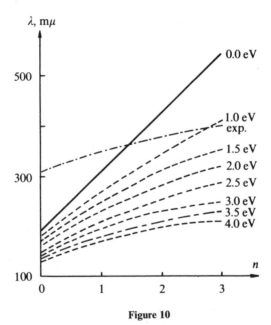

Figure 10

Experimental and theoretical curves of λ_{max} of 5-nitrofuryl-2-polyenealdehydes as a function of the number of conjugated C=C bonds in the side chain.

can be made to fit the experimental data if a correction term A is added:

$$\lambda_1 = \left[\frac{V_0}{hc}\left(1 - \frac{1}{N}\right) + \frac{h}{8mcL^2}(N + 1) \right]^{-1} + A. \qquad (3)$$

We thus obtain, for 5-nitrofuryl-2-polyeneketones and 5-nitrofuryl-2-polyenealdehydes (Figures 9 and 10)

$$\lambda_1^{ket} = \left[\frac{2.2}{hc} \left(1 - \frac{1}{N} \right) + \frac{h}{8mcL^2} (N + 1) \right]^{-1} + 75 \, m\mu$$

and

$$\lambda_1^{ald} = \left[\frac{3.5}{hc} \left(1 - \frac{1}{N} \right) + \frac{h}{8mcL^2} (N + 1) \right]^{-1} + 165 \, m\mu.$$

In order to calculate the variation in the maxima of the two absorption bands, at least two experimental points are required for each band for the determination of the slope of the curve-fitting line (for more accurate calculations, more than two experimental points are needed), whence the optimum values of the parameters V_0 and A are obtained. Since each 5-nitrofuryl-2-polyene series comprises only a few members, the value of the method for calculating the spectra of these compounds is highly limited.

Another major disadvantage of the method is that only the first absorption band (the one with the longer wavelength) can be employed, i.e., only the first electronic transition is used. Attempts to calculate the second electronic transition failed, and the results differed considerably from the true situation.

Generally speaking, Kuhn's model of a free electron in a potential well is applicable, from the physical point of view, to conjugated systems of π electrons of this type. It is therefore reasonable to try to find another function for the potential distribution at the bottom of the well which would give a better fit to the energy levels of these systems, while retaining Kuhn's basic approach.

Thus, for instance, the potential field of the π electrons in a polyene molecule can be approximated with a one-dimensional potential well, whose width and depth are functions of the "length" of the molecule. A suitable potential is

$$U(x) = - U_0 (\tanh^2 (x/d) - 1), \tag{4}$$

where U_0 is the maximum depth of the well and d is the effective "width" (more exactly, the width between $U = U_0$ and $U = U_0/e$). The shape of the potential well is shown in Figure 11.

35

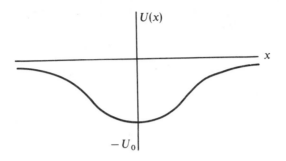

Figure 11

Shape of potential well for a polyene molecule.

This potential is consistent with general physical considerations (provided an allowance is made for the potential change at the end groups), and is also convenient because the Schroedinger equation has an exact solution in this case. It gives the following expression for the energy levels [83, 84]:

$$E_n = \frac{1}{2d^2} \left\{ 2U_0 d^2 - \left[\sqrt{2U_0 d^2 + \tfrac{1}{4}} - (n + \tfrac{1}{2}) \right]^2 \right\}, \tag{5}$$

where $n = 0, 1, 2, 3 \ldots$

In comparing the wavelengths of the electronic transitions obtained in this model with their experimental values, it must be borne in mind that the transitions from E_0 to E_2, E_4, etc. (i.e., to even levels) are forbidden by the selection rules [83]. Thus, if we denote by λ_1 and λ_2 the wavelengths of the first and the second allowed transition (i.e., from E_0 to E_1 and to E_3), we obtain the expressions

$$\lambda_1 = \frac{2\pi c d^2}{\sigma - 1} \quad \text{and} \quad \lambda_2 = \frac{2\pi c d^2}{3(\sigma - 3)},$$

where

$$\sigma = \sqrt{2U_0 d^2 + \tfrac{1}{4}}.$$

Calculations show that it is possible to select such well parameters and such values of d^2 and $U_0 d^2$ for the given length of the molecule that the values of λ_1 and λ_2 obtained for a large sample of molecules of various lengths are close to the experimental values.

36

Let us examine the series of 5-nitrofuryl-2-polyeneketones, as before. The "length" of the molecule in this context is again described by the number of π electrons, N. Experimental data are available for $N = 4, 6, 8, 10, 12$. For molecules of this type we have to adopt the following dependence of d^2 on U and N:

1) d^2 is a linear function of N:

$$d^2 = a^2(N + B),$$

2) $U_0 d^2$ is independent of N, i.e.,

$$U_0 d^2 = \text{const.}$$

In other words, the width of the well increases with the length of molecule (the width squared varies linearly with N), while the depth of the well decreases so that $U_0 d^2$ remains constant.

Thus, the three parameters, a^2, B, and $U_0 d^2 = y/2$, must be determined experimentally for the wavelengths corresponding to the absorption bands maxima, and then the absorption maxima of the other molecules are calculated (Table 6). The following parameters were determined experimentally:

$$\lambda_{1 \text{ exp}}(N = 4) = 299 \text{ m}\mu,$$

$$\lambda_{2 \text{ exp}}(N = 4) = 221 \text{ m}\mu,$$

$$\lambda_{1 \text{ exp}}(N = 12) = 430 \text{ m}\mu,$$

Table 6
MAXIMA OF THE ABSORPTION BANDS
OF 5-NITROFURYL-2-POLYENEKETONES

n	N	Calculated $\lambda_{1 \text{ max}}$, mμ	Experimental $\lambda_{1 \text{ max}}$, mμ	Calculated $\lambda_{2 \text{ max}}$, mμ	Experimental $\lambda_{2 \text{ max}}$, mμ
0	4	299*	299	221*	221
1	6	332	340	245	239
2	8	364	377	269	274
3	10	397	403	293	305
4	12	430*	430	318	330

* Experimental values.

whence

$$(a')^2 = 2\pi c a^2 = 29.83,$$
$$B = 14.26,$$
$$y = 2U_0 d^2 = 7.714.$$

The results of the calculations are shown in Table 6.

The wavelengths of the absorption bands for other 5-nitrofuryl-2-polyenes were calculated in a similar manner. The following experimental points were chosen for the calculations in the case of 5-nitrofuryl-2-polyene-aldehydes:

$$\lambda_{1 \ exp} (N = 4) \ = 310 \ m\mu,$$
$$\lambda_{2 \ exp} (N = 4) \ = 226 \ m\mu,$$
$$\lambda_{1 \ exp} (N = 10) = 399 \ m\mu,$$

whence

$$(a')^2 = 54.72,$$
$$B = 16.89,$$
$$y = 7.733.$$

The results of the calculations are shown in Table 7.

Table 7

MAXIMA OF THE ABSORPTION BANDS
OF 5-NITROFURYL-2-POLYENEALDEHYDES

n	N	Calculated $\lambda_{1 \ max}$, $m\mu$	Experimental $\lambda_{1 \ max}$, $m\mu$	Calculated $\lambda_{2 \ max}$, $m\mu$	Experimental $\lambda_{2 \ max}$, $m\mu$
0	4	310*	310	226*	226
1	6	340	345	247	240
2	8	369	377	269	270
3	10	399*	399	291	304

* Experimental values.

The differences between the calculated and the experimental λ_{max} do not exceed 4%. Unlike Kuhn's method, this technique also gives the wavelength of the second absorption band as well, which corresponds to the transition between the ground level and the third electronic level. The agreement with the experimental data is much closer than in Kuhn's method. Three experimental points are required, as compared to Kuhn's two, but both absorption bands can be calculated, and not only one.

The third allowed transition, which produces the third absorption band, cannot be calculated, since for the third transition

$$\lambda_3 = \frac{2\pi c d^2}{5(\sigma - 5)}.$$

For the molecules under consideration $\sigma = 4.643$ and 4.686, respectively, which means that λ_3 is negative. The physical meaning of this is that the potential well chosen as a model for the molecules is too shallow to accommodate the fifth electronic level, which is responsible for the third absorption band. Nevertheless, such bands do in fact appear in the spectra of the higher 5-nitrofuryl-2-polyenes; thus, in 5-nitrofuryl-2-polyeneketones for $n = 3$ we have $\lambda_3 = 236$ mμ, and for $n = 4$, $\lambda_3 = 244$ mμ.

Thus, the use of the potential well model leads to a fairly satisfactory fit with the experimental data. However, the model ignores the asymmetry of the molecule. The second drawback is that the potential well is too shallow. The former shortcoming may be rectified by using an asymmetric model, such as that proposed by Morse and Feshbach [83], but the calculations employing this model gave only slightly better results for two absorption bands, while requiring yet another experimental point. The third electronic transition could not be calculated since the well again was too shallow. Obviously, this question requires further study.

2. The LCAO Approximation of the Method of Molecular Orbitals

This method, which was first used by Mulliken [85] and subsequently developed by Herzberg [86], Sklar [87], Lennard-Jones [88] and others, is now frequently employed. It should be noted at the outset that the method, just like the previous one, is semi-empirical, since it involves the use of suitably chosen starting parameters. Nevertheless, the results obtained are highly valuable and interesting and we have attempted to apply it to the systems which constitute the subject of this book.

The furan molecule is the starting molecule in the calculations; this is a five-center system which can also be regarded as a π-electron sextet

39

(including the lone pair electrons of the oxygen atom). The presence of the lone pair electrons should be reflected in the values of the Coulomb and the resonance integrals (i.e., in the values of α_{11} and $\beta_{12} = \beta_{15}$).

The initial data for the calculation of the furan molecule were borrowed from Roberts [89] and from Gren (private communication). The calculations are based on the following data:

$$\alpha_{11} = \alpha_0 + 2\beta_0; \qquad \beta_{12} = \beta_{15} = 0.98\beta_0;$$
$$\alpha_{22} = \alpha_{55} = \alpha_0 + 0.2\beta_0; \qquad \beta_{23} = \beta_{45} = 1.1\beta_0;$$
$$\alpha_{33} = \alpha_{44} = \alpha_0. \qquad \beta_{13} = 0.8\beta_0.$$

The energy values are written in the form

$$E = \alpha_0 + x\beta_0.$$

The calculations were performed by R. Kalnyn' in the Computation Center of the Latvian State University. The values of x obtained for the 5-nitrofuryl-2-polyenealdehyde series are shown in Table 8. The values of β_0 were found from spectroscopic data, which give $\beta_{0av} = 71.9$ kcal/mole.

The energy level diagram for 5-nitrofuryl-2-polyenealdehydes, showing the allowed transitions, is given in Figure 12 (experimental values in brackets).

Table 9 lists the calculated and the experimental wavelengths in the absorption spectra of 5-nitrofuryl-2-polyenealdehydes.

The average variance between the calculated and the experimental data is 13%.

We also calculated the electron densities for the molecules of furan, nitrofuran, and 5-nitrofurfural (Table 10).

The electron density for the furan molecule was calculated by various authors; a recent calculation, in which a cumbersome approximation method was employed, was carried out by Sappenfield and Kreevoy [90], and the results obtained differ by a mere 2–6% from those of Table 10.

The relatively high charge density on the C_2 and C_5 atoms in furan is in good agreement with the fact that electrophilic substitution of furan preferentially takes place in these positions, rather than at C_3 and C_4. The high charge density on the oxygen atom is striking. The charge delocalization does not exceed 10% (see discussion of the aromatic nature of furan).

VALUES OF THE COEFFICIENTS x FOR THE CALCULATION OF THE MOLECULES OF FURAN AND A NUMBER OF NITROFURANS ACCORDING TO THE MO LCAO METHOD

Compound	Coefficient x				
Furan (ring: O-1, 2, 3, 4, 5)	−2.693	−1.240	−0.908	+0.934	+1.503
Nitrofuran (ring O-1, 2, 3, 4, 5, N-6, O-7, O-8)	−3.023 +0.299	−2.544 +1.111	−1.245 +1.576	−1.200 —	−0.971 —
(ring with CH=O 9,10; O-8, N-3)	−3.040 −1.021	−2.636 +0.219	−2.221 +0.675	−1.252 +1.334	−1.200 +1.641
(CH=CH—CH=O 9,10,11,12)	−3.031 −1.200 +1.134	−2.582 −0.783 +1.750	−2.315 +0.236 —	−1.441 +0.552 —	−1.212 +1.492 —
(CH=CH—CH=CH—CH=O 8,9,10,11,12,13)	−3.031 −1.200 +0.956	−2.580 −1.129 +1.352	−2.322 −0.653 +1.583	−1.641 +0.242 +1.801	−1.261 +0.482 —
(CH=CH—CH=CH—CH=CH—CH=O 9,10,11,12,13,14,15,16)	−3.023 −1.242 +0.403 +1.777	−2.544 −1.200 +0.921 —	−2.322 −0.971 +1.111 —	−1.710 −0.655 +1.429 —	−1.248 +0.299 +1.576 —

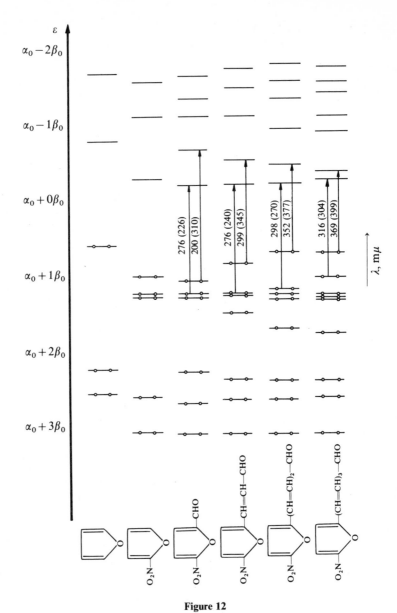

Figure 12

Energy levels of furan and some furan derivatives.

Table 9

MAXIMA OF THE ABSORPTION BANDS OF 5-NITROFURYL-2-
POLYENEALDEHYDES CALCULATED BY THE MO LCAO METHOD

n	Calculated $\lambda_{1\,max}$, mμ	Experimental $\lambda_{1\,max}$, mμ	Calculated $\lambda_{2\,max}$, mμ	Experimental $\lambda_{2\,max}$, mμ
0	260	310	278	226
1	299	345	276	240
2	252	377	298	270
3	362	399	316	304

Table 10

CHARGE DENSITY DISTRIBUTION IN MOLECULES OF FURAN,
2-NITROFURAN, AND 5-NITROFURFURAL

Compound			
Electron density	$q_1 = 1.808$ $q_2 = 1.067$ $q_3 = 1.030$ $q_4 = 1.030$ $q_5 = 1.067$	$q_1 = 1.808$ $q_2 = 1.002$ $q_3 = 1.030$ $q_4 = 0.750$ $q_5 = 1.192$ $q_6 = 1.393$ $q_7 = 1.575$ $q_8 = 1.575$	$q_1 = 1.811$ $q_2 = 1.066$ $q_3 = 0.929$ $q_4 = 0.928$ $q_5 = 1.056$ $q_6 = 0.987$ $q_7 = 1.569$ $q_8 = 1.569$ $q_9 = 1.185$ $q_{10} = 1.367$

The appearance of a nitro group results in a strong shift of the electron cloud (Table 10), mainly away from C_4, and in an increased electron density on C_5. If the electron-accepting carbonyl group is introduced as a substituent at C_2, the electron density distribution is leveled out to some extent: it again increases on C_4 and decreases on C_5. This is a confirmation of the ideas discussed above regarding the competing effects of the 5-nitro substituents and of the 2-carbonyl substituent on the electron cloud of the entire molecule.

43

The results of the theoretical calculations are consistent with the interpretation of the electronic absorption bands as being due to transitions within a single system of π electrons (see above).

Moreover, it is also confirmed that the aromatic nature of furan is weak (see above), since the system of —C$=$C— bonds in the furan ring is treated in the calculations as a system of conjugated bonds in an aliphatic diene, and the results obtained in this approximation agree with the experimental data.

PART II

Atlas of Ultraviolet Absorption Spectra
of 5-Nitrofurans

Explanations to the Atlas

The electronic spectra of 5-nitrofurans were taken using an SF-4 spectrophotometer. The compounds were recrystallized three times before the spectrum was taken. The melting points and the literature references for the preparation and purification of the compounds are given in the legends to each figure. The legend also contains the chemical name of the compound, its empirical formula, and molecular weight calculated on the oxygen (chemical) scale from the international atomic weights of 1957, which were corrected in 1960 by the International Committee on Atomic Weights.

The spectra were taken in duplicate on solutions of two weighed samples, which were dissolved right before taking the spectrum. The solvent was ethanol, purified in the usual way [91].

The electronic absorption curves of 5-nitrofurans are given in Figures 13–62. The wavelengths are given in millimicrons (1 mμ = 10^{-9} m). The corresponding frequencies in cm^{-1} (v) are indicated on the top of each figure.

The spectra are given both on the linear (a) and on the logarithmic (b) scale.

In order to enable the reader to make a proper evaluation of the intensity and of the shape of the absorption curves, λ_{max}, ε, log ε, and the spectral bandwidth (cm^{-1}) are all shown.

47

2-NITROFURAN

Figure 13

2-Nitrofuran [92].

M. wt. 113.08; $C_4H_3NO_3$; m. p. 28–29°.

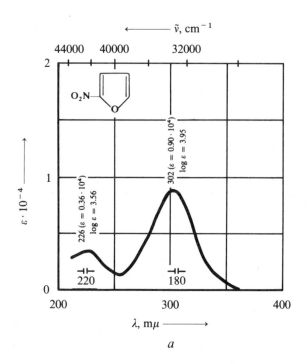

a

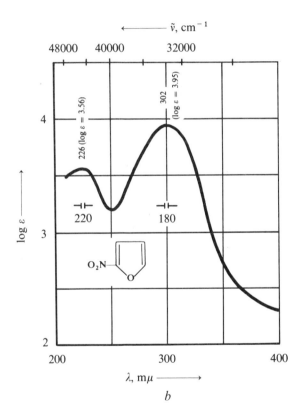

b

5-NITRO-2-METHYLFURAN

Figure 14

5-Nitro-2-methylfuran [93].

M. wt. 127.10; $C_5H_5NO_3$; m. p. 41–42°.

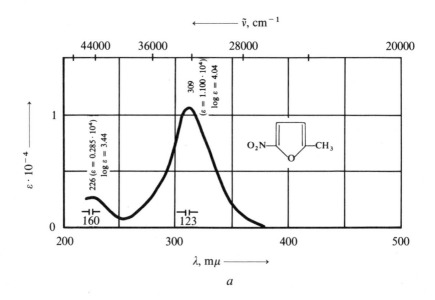

a

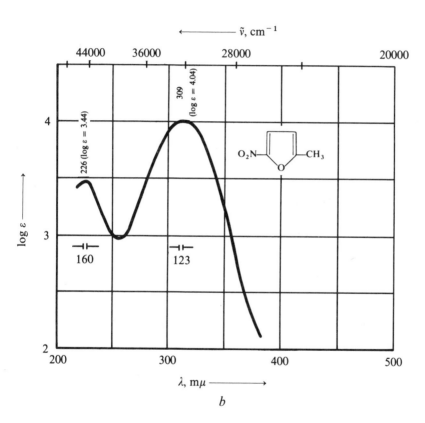

$\tilde{v}$, cm^{-1}

44000   36000   28000   20000

309 (log ε = 4.04)

226 (log ε = 3.44)

O$_2$N⎯〈 〉⎯CH$_3$

160   123

λ, mμ ⎯⎯→

b

ALDEHYDES OF THE 5-NITROFURAN SERIES

Figure 15

5-Nitrofurfural [94].

M. wt. 141.09; $C_5H_3NO_4$; m. p. 35–36°.

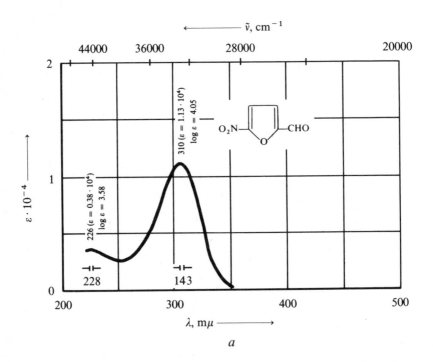

a

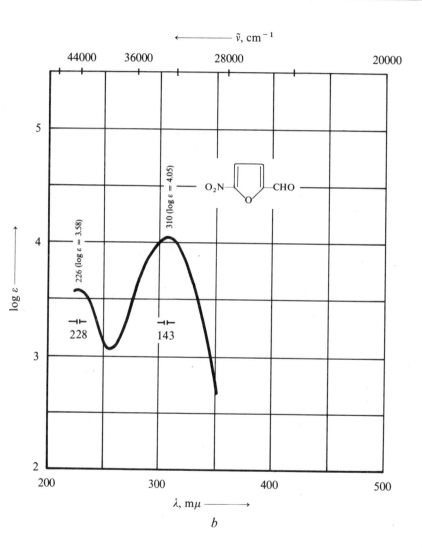

$\tilde{v}$, cm^{-1}

b

ALDEHYDES OF THE 5-NITROFURAN SERIES

Figure 16

β-(5-Nitrofuryl-2)-acrolein [95, 96].
M. wt. 167.13; $C_7H_5NO_4$; m. p. 117–118°.

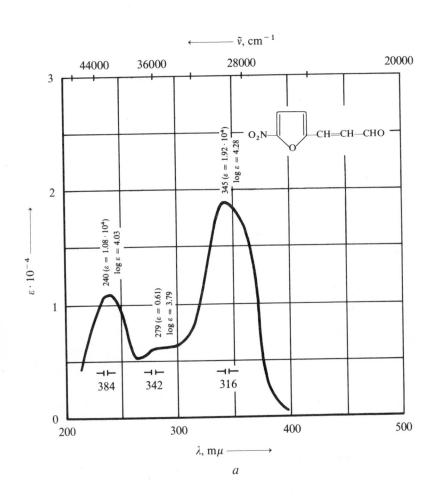

a

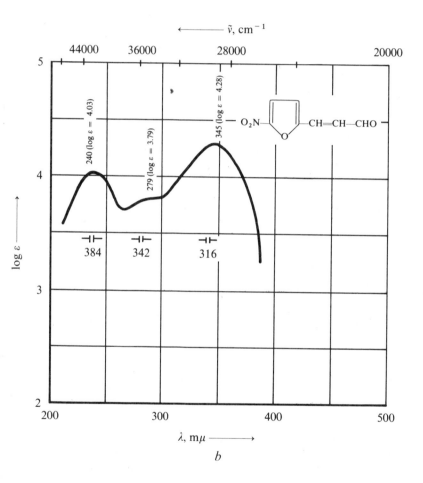

b

Figure 17

5-(5'-Nitrofuryl-2')-pentadien-2,4-al-1 [62].

M. wt. 193.16; $C_9H_7NO_4$; m. p. 125–126°.

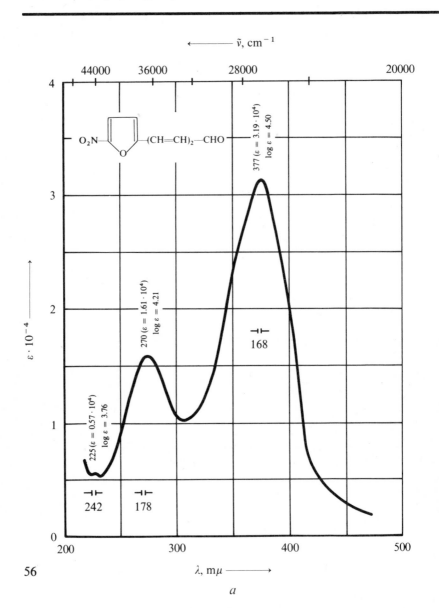

a

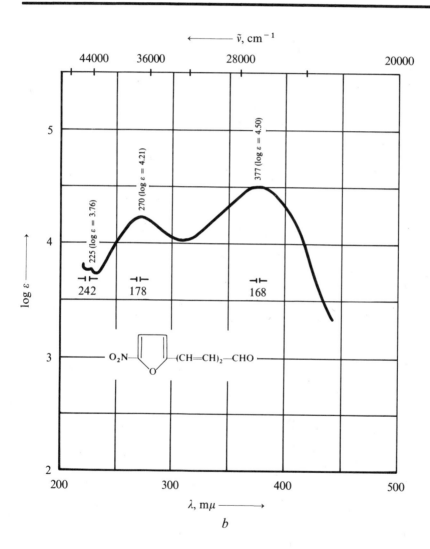

b

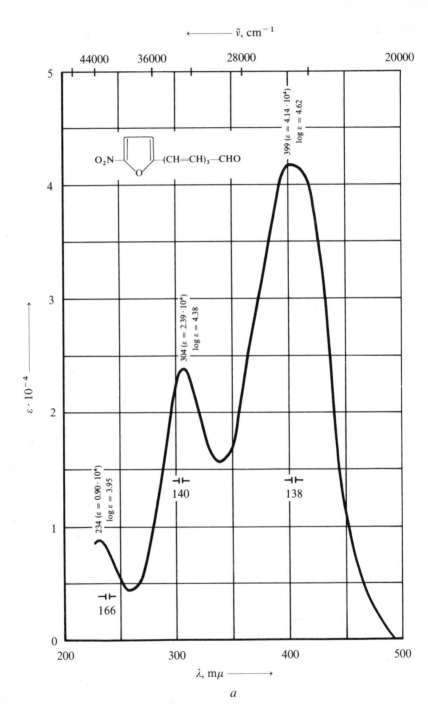

a

ALDEHYDES OF THE 5-NITROFURAN SERIES

Figure 18

7-(5'-Nitrofuryl-2')-heptatrien-2,4,6-al-1 [97].
M. wt. 219.20; $C_{11}H_9NO_4$; m. p. 148–150°.

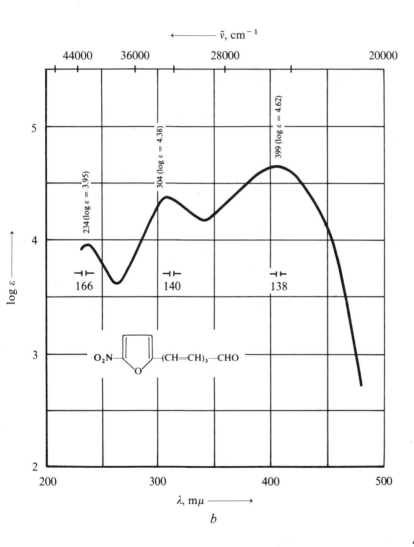

b

KETONES OF THE 5-NITROFURAN SERIES

Figure 19

5-Nitro-2-acetylfuran [98].
M. wt. 155.11; $C_6H_5NO_4$; m. p. 78°.

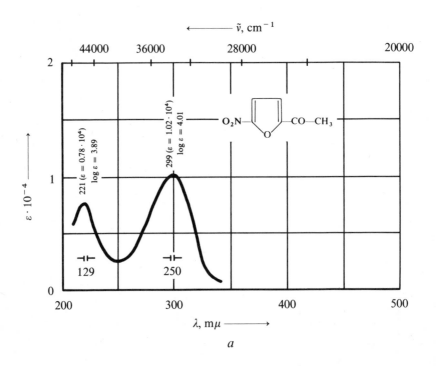

a

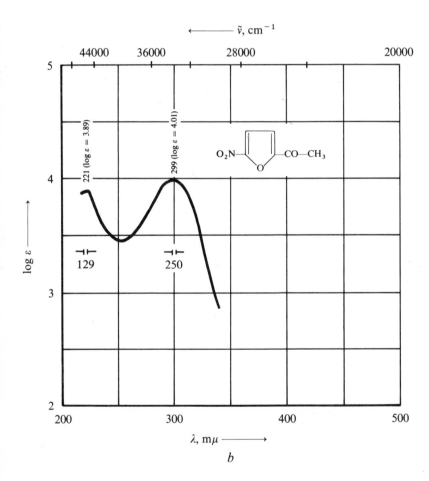

b

Figure 20

1-(5'-Nitrofuryl-2')-buten-1-one-3 [99].
M. wt. 181.15; $C_8H_7NO_4$; m. p. 114°.

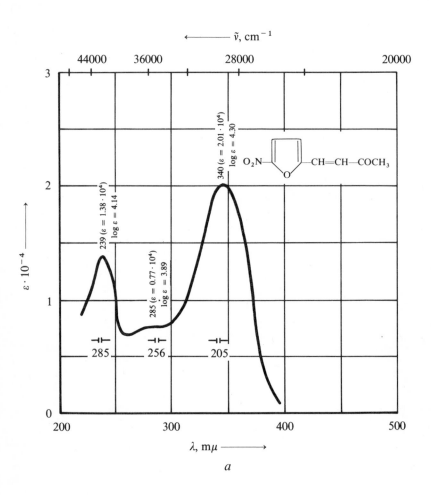

a

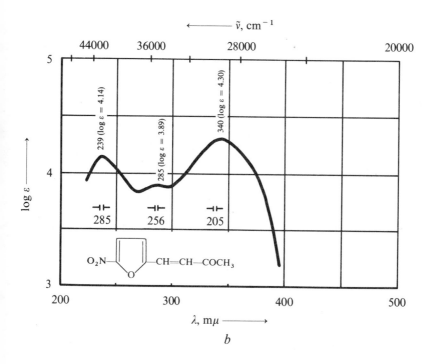

$$b$$

KETONES OF THE 5-NITROFURAN SERIES

Figure 21

1-(5'-Nitrofuryl-2')-hexadien-1,3-one-5 [100].
M. wt. 207.19; $C_{10}H_9NO_4$; m. p. 136°.

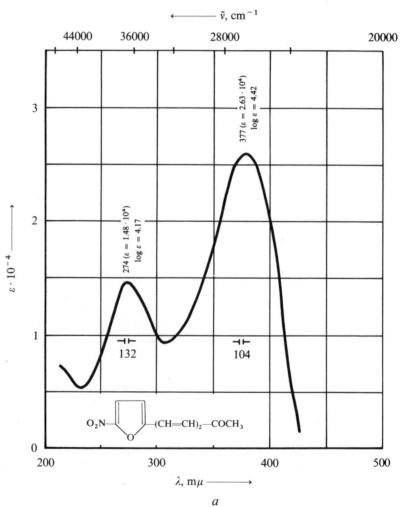

a

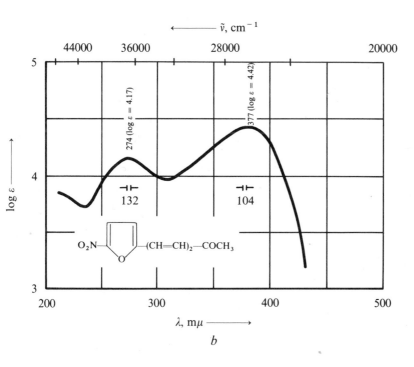

b

Figure 22

1-(5′-Nitrofuryl-2′)-octatrien-1,3,5-one-7 [100].
M. wt. 233.23; $C_{12}H_{11}NO_4$; m. p. 138°.

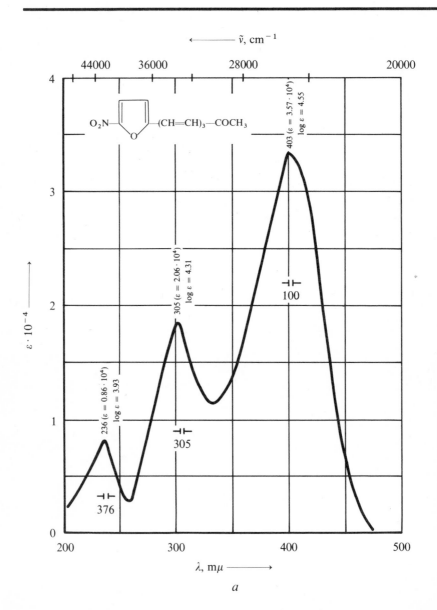

a

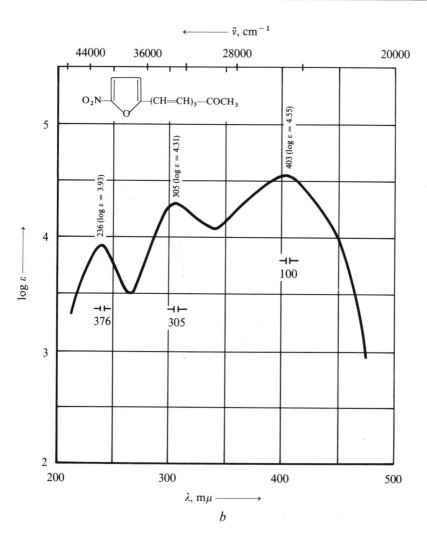

b

67

Figure 23

1-(5′-Nitrofuryl-2′)-decatetraen-1,3,5,7-one-9 [100].
M. wt. 259.27; $C_{14}H_{13}NO_4$; m. p. 160°.

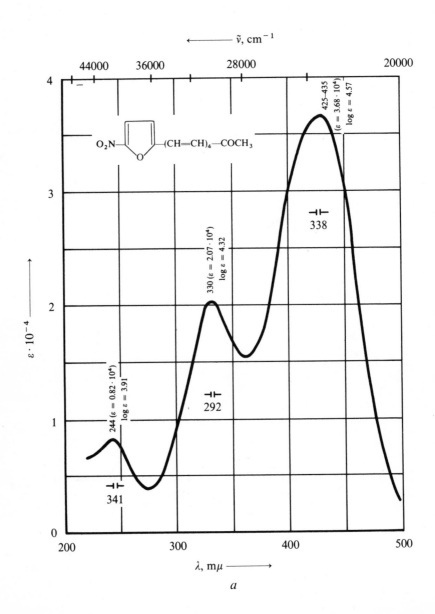

a

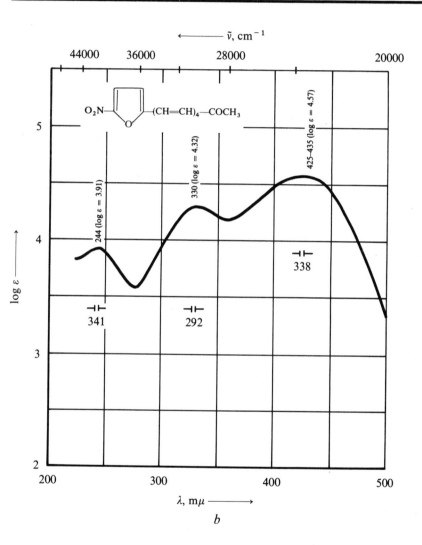

b

CARBOXYLIC ACIDS OF THE 5-NITROFURAN SERIES

Figure 24

5-Nitro-2-furoic acid [101].

M. wt. 157.09; $C_5H_3NO_5$; m. p. 185°.

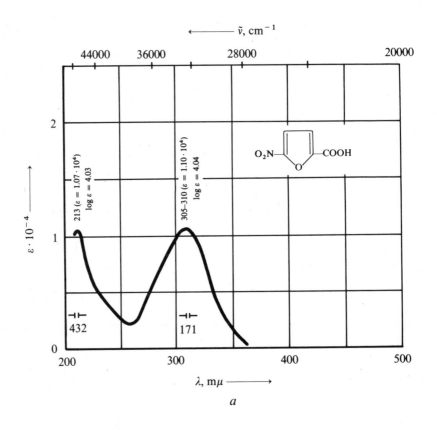

a

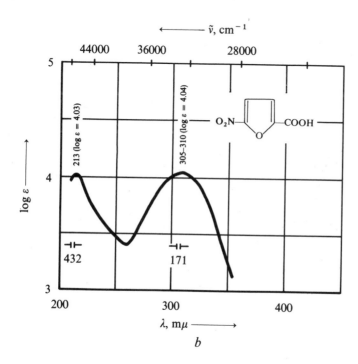

b

CARBOXYLIC ACIDS OF THE 5-NITROFURAN SERIES

Figure 25

β-(5-Nitrofuryl-2)-acrylic acid [102].

M. wt. 183.13; $C_7H_5NO_5$; m. p. 234–236° (decomp.).

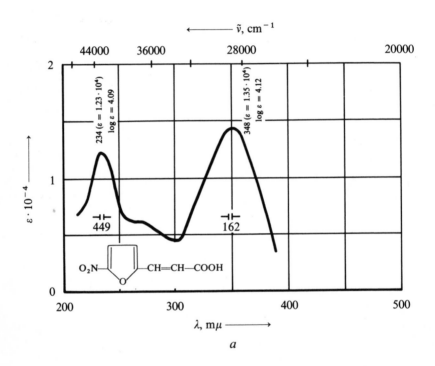

a

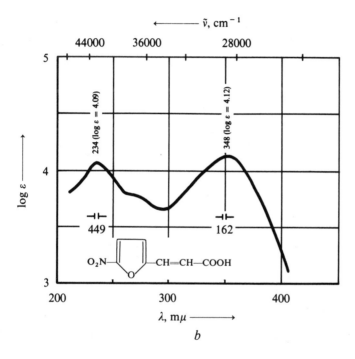

b

ESTERS OF CARBOXYLIC ACIDS OF THE 5-NITROFURAN SERIES

Figure 26

5-Nitro-2-furoic acid methyl ester [103].

M. wt. 171.11; $C_6H_5NO_5$; m. p. 80–81.5°.

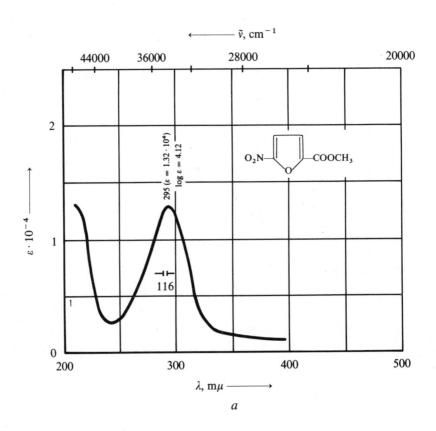

a

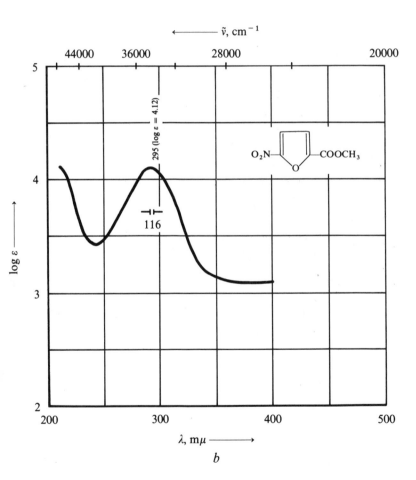

b

75

ESTERS OF CARBOXYLIC ACIDS OF THE 5-NITROFURAN SERIES

Figure 27

5-Nitro-2-furoic acid ethyl ester [104, 105].
M. wt. 185.14; $C_7H_7NO_5$; m. p. 100–101°.

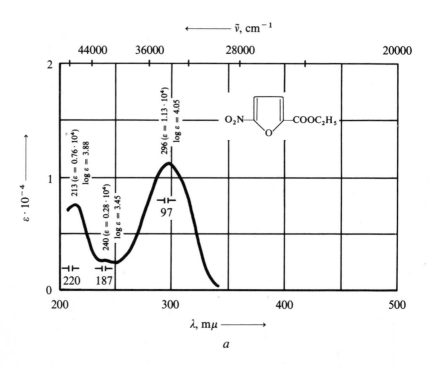

a

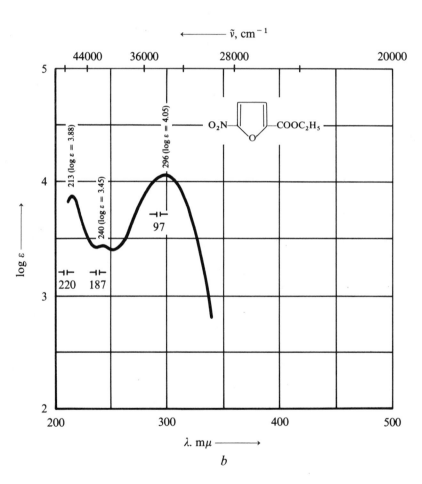

b

Figure 28

β-(5-Nitrofuryl-2)-acrylic acid ethyl ester [94].

M. wt. 211.18; $C_9H_9NO_5$; m. p. 123°.

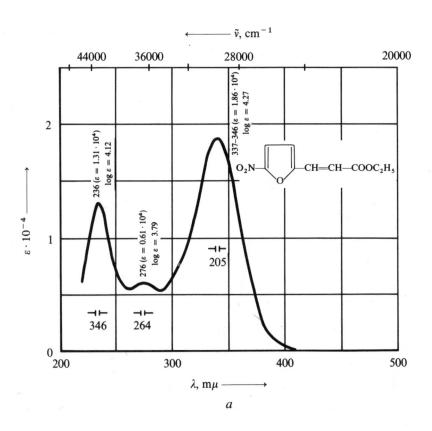

a

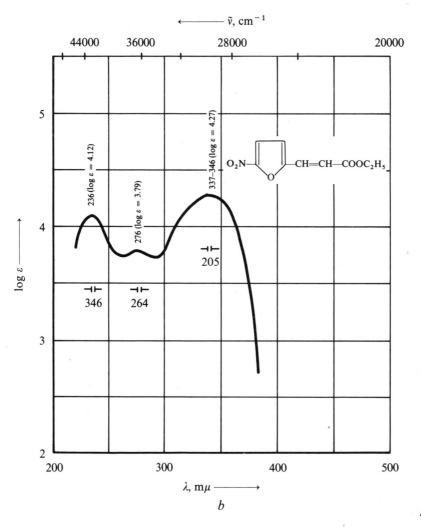

b

79

ESTERS OF CARBOXYLIC ACIDS OF THE 5-NITROFURAN SERIES

Figure 29

5-(5'-Nitrofuryl-2')-pentadiene-2,4-carboxylic acid ethyl ester [100].
M. wt. 237.22; $C_{11}H_{11}NO_5$; m. p. 122–123°.

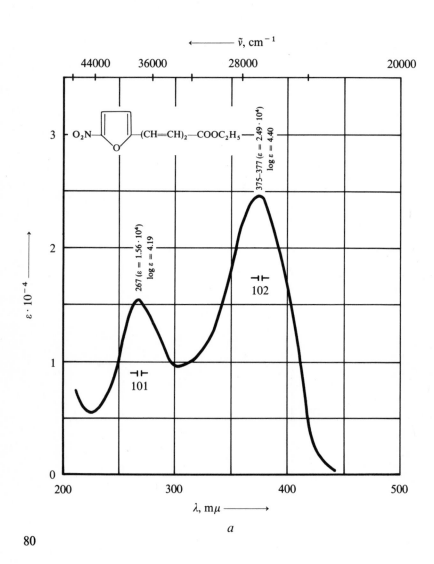

a

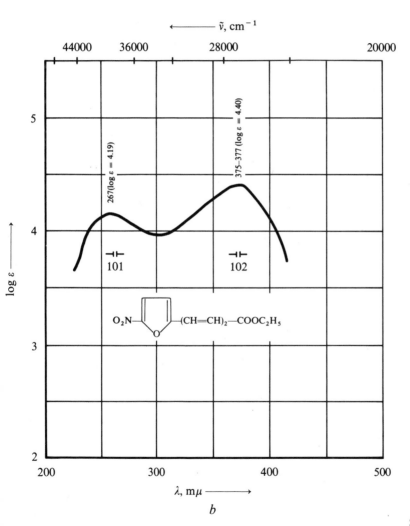

b

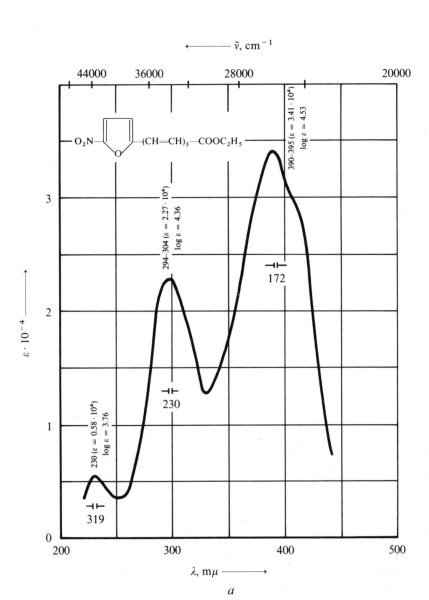

a

ESTERS OF CARBOXYLIC ACIDS OF THE 5-NITROFURAN SERIES

Figure 30

7-(5'-Nitrofuryl-2')-heptatriene-2,4,6-carboxylic acid ethyl ester [100].
M. wt. 263.26; $C_{13}H_{13}NO_5$; m. p. 137–138.5°.

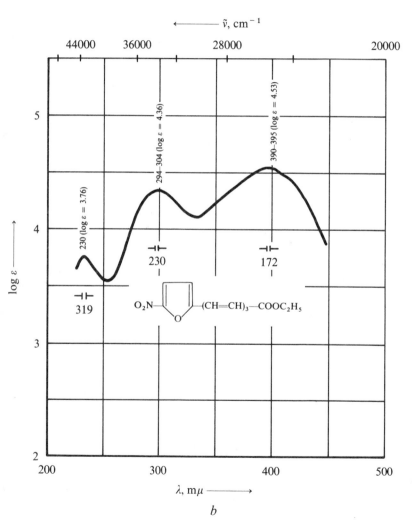

b

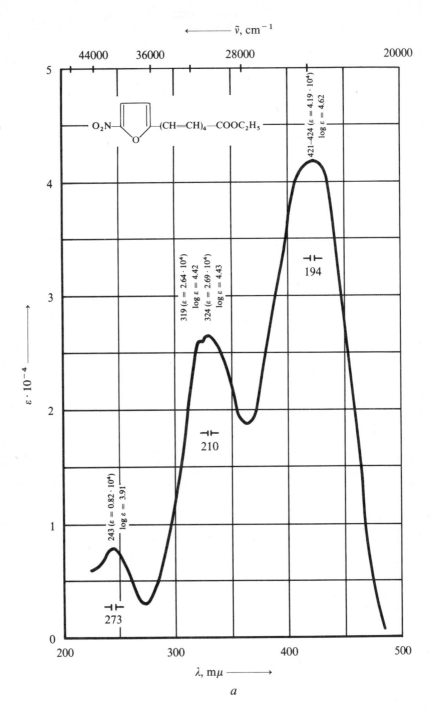

a

ESTERS OF CARBOXYLIC ACIDS OF THE 5-NITROFURAN SERIES

Figure 31

9-(5′-Nitrofuryl-2′)-nonatetraene-2,4,6,8-carboxylic acid ethyl ester [100].
M. wt. 289.29; $C_{15}H_{15}NO_5$; m. p. 156–157°.

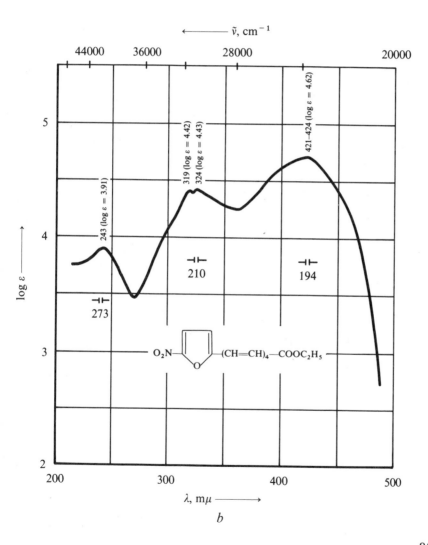

b

GLYCOLIC ACETATES

Figure 32

5-Nitrofurfural diacetate [94].

M. wt. 243.18; $C_9H_9NO_7$; m. p. 92°.

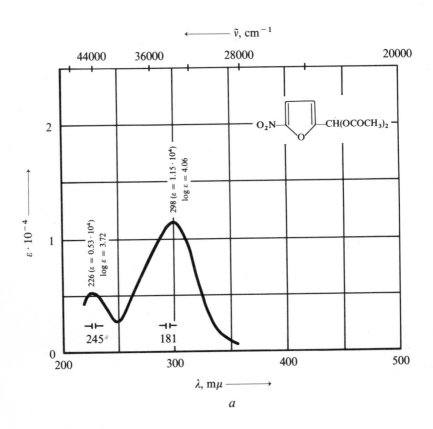

a

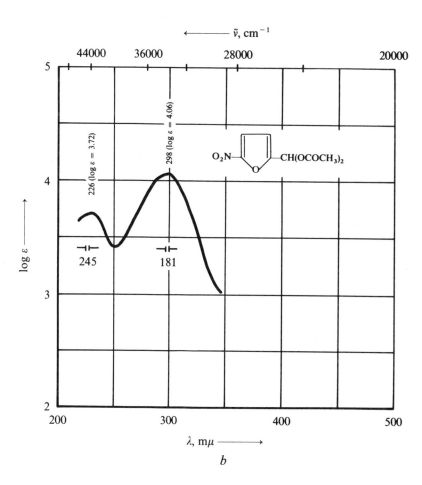

b

GLYCOLIC ACETATES

Figure 33

β-(5-Nitrofuryl-2)-acrolein diacetate [96].

M. wt. 269.22; $C_{11}H_{11}NO_7$; m. p. 97–98°.

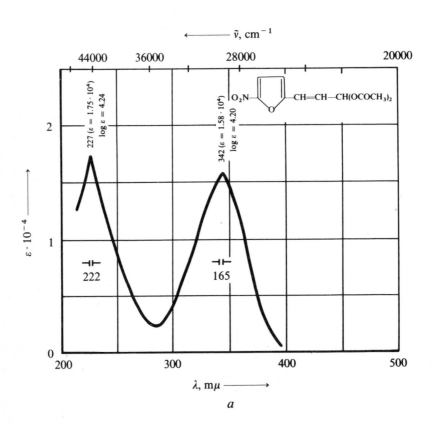

a

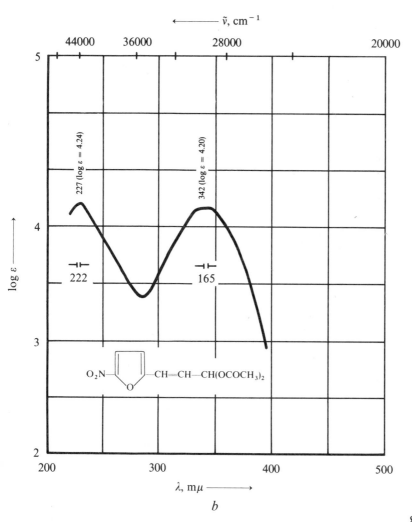

b

89

GLYCOLIC ACETATES

Figure 34

5-(5′-Nitrofuryl-2′)-pentadien-2,4-al-1-diacetate [96].
M. wt. 295.26; $C_{13}H_{13}NO_7$; m. p. 118–119°.

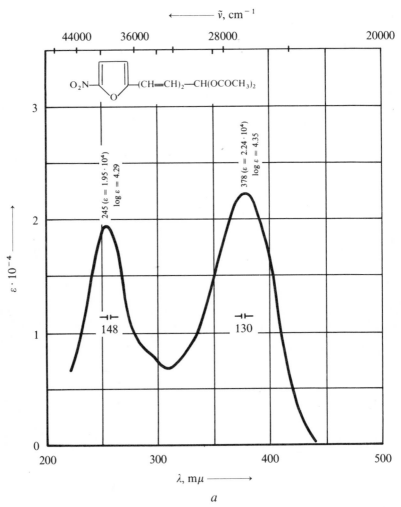

a

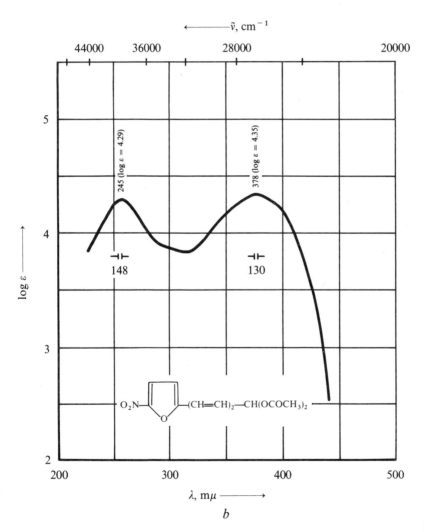

b

GLYCOLIC ACETATES

Figure 35

7-(5'Nitrofuryl-2')-heptatrien-2,4,6-al-1-diacetate.
Obtained by a method analogous to that in [96].
M. wt. 321.29; $C_{15}H_{15}NO_7$; m. p. 135°.

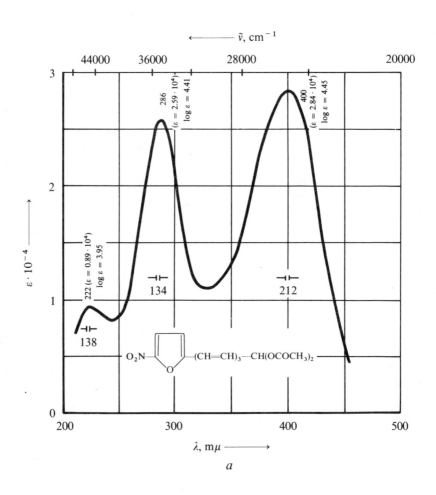

a

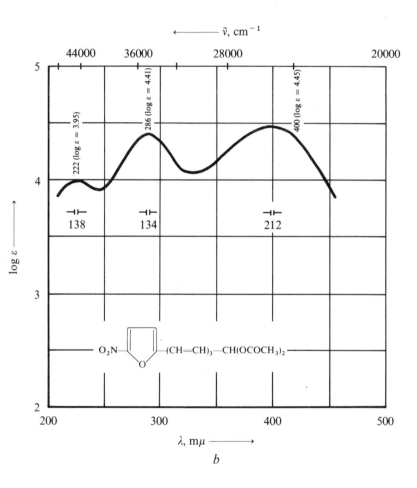

b

SEMICARBAZONES

Figure 36

5-Nitro-2-furfurylidene semicarbazone [106].
M. wt. 198.15; $C_6H_6N_4O_4$; m. p. 236–240° (decomp.).

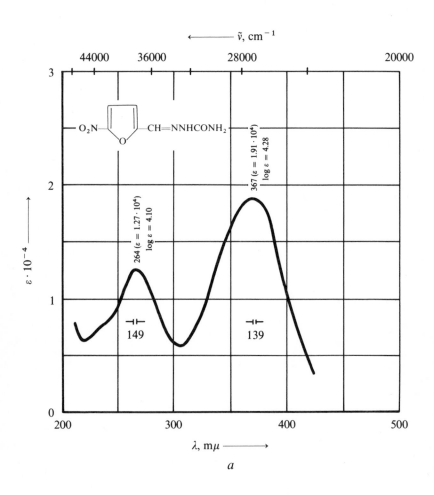

a

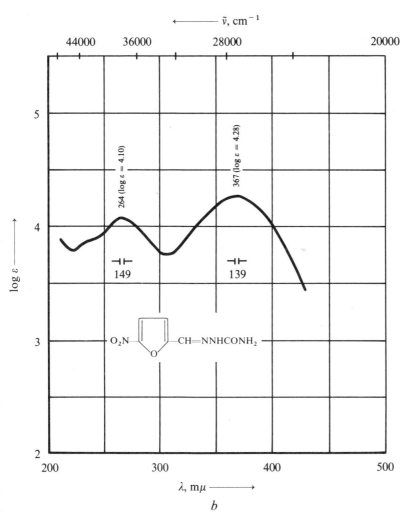

ṽ, cm⁻¹ is labeled as $\tilde{v}$, cm^{-1}

264 (log ε = 4.10)

367 (log ε = 4.28)

149

139

O_2N——CH=NNHCONH$_2$

b

95

Figure 37

β-(5-Nitrofuryl-2)-acrylidene semicarbazone [107].

M. wt. 224.18; $C_8H_8N_4O_4$; m. p. 240–242° (decomp.).

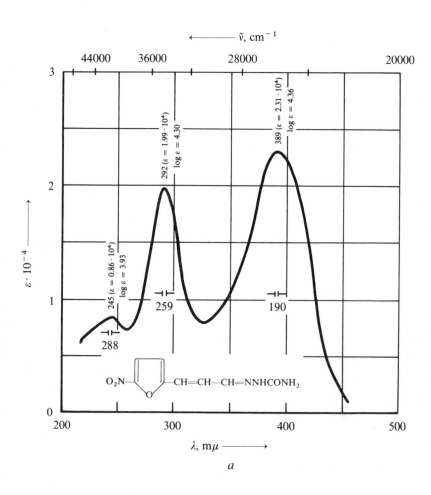

a

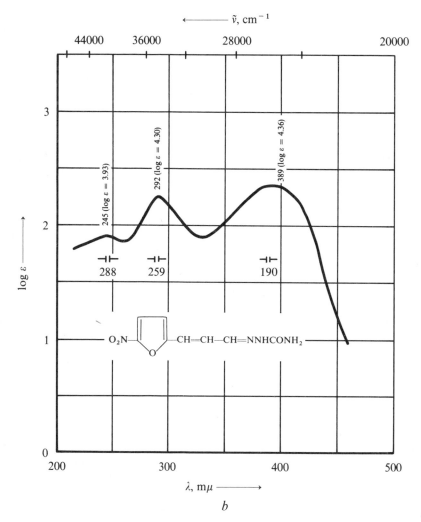

b

Figure 38

5-(5'-Nitrofuryl-2')-pentadien-2,4-al-1-semicarbazone [62].

.M. wt. 250.22; $C_{10}H_{10}N_4O_4$; m. p. 240° (decomp.).

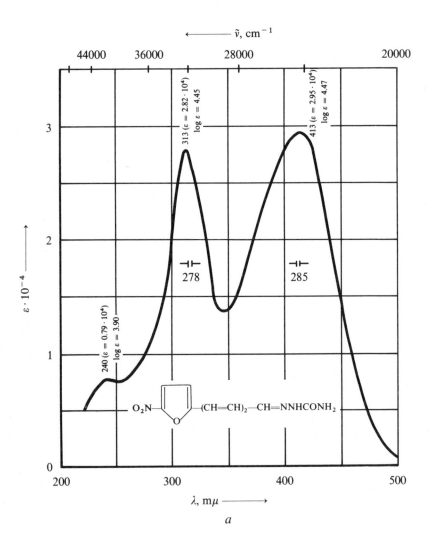

a

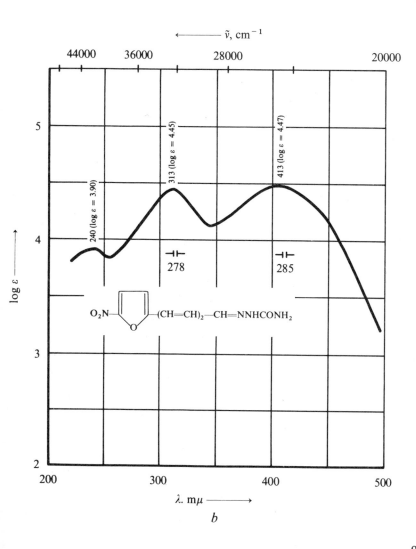

b

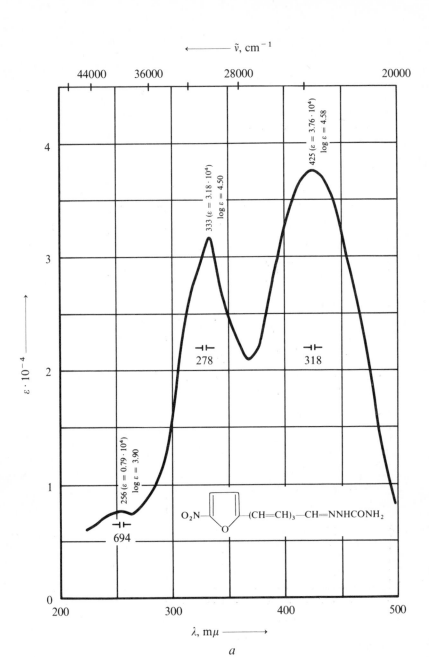

a

SEMICARBAZONES

Figure 39

7-(5'-Nitrofuryl-2')-heptatrien-2,4,6-al-1-semicarbazone [62].
M. wt. 276.26; $C_{12}H_{12}N_4O_4$; m. p. 250° (decomp.).

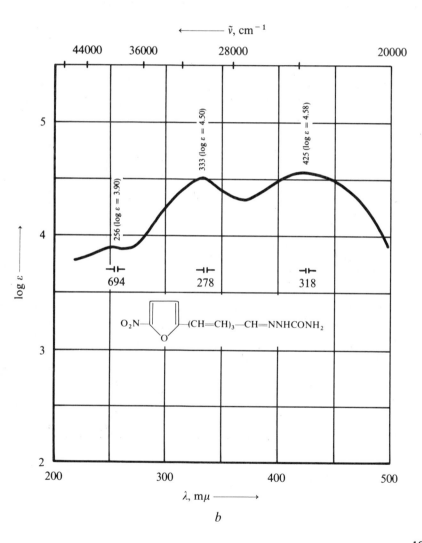

b

THIOSEMICARBAZONES

Figure 40

5-Nitro-2-furfurylidene thiosemicarbazone [106].
M. wt. 214.21; $C_6H_6N_4O_3S$; m. p. 232–235° (decomp.).

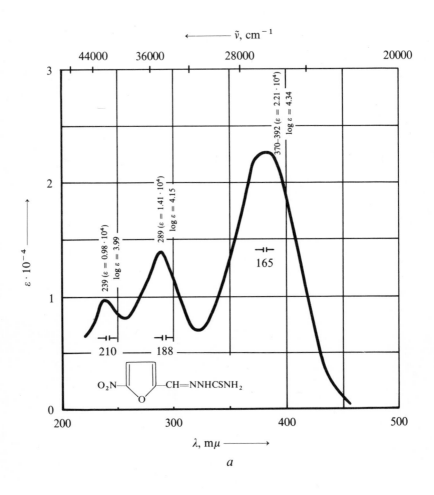

a

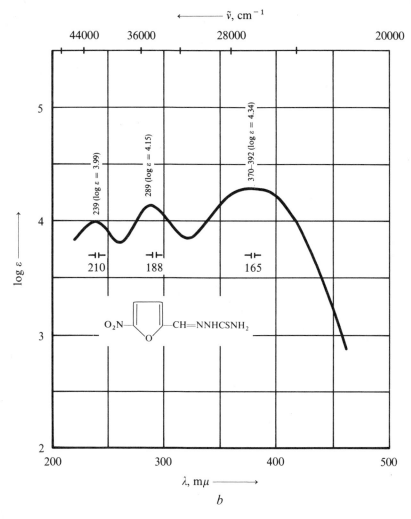

b

THIOSEMICARBAZONES

Figure 41

β-(5-Nitrofuryl-2)-acrylidene thiosemicarbazone [108].
M. wt. 240.25; $C_8H_8N_4O_3S$; m. p. 221° (decomp.).

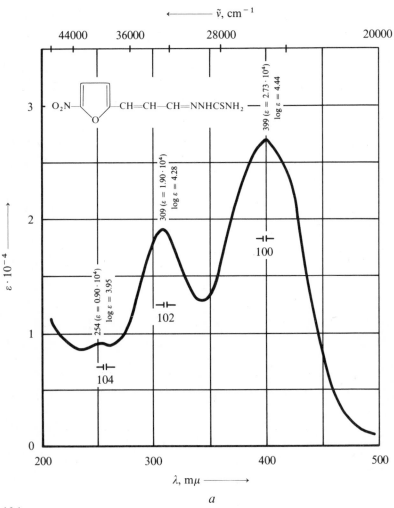

a

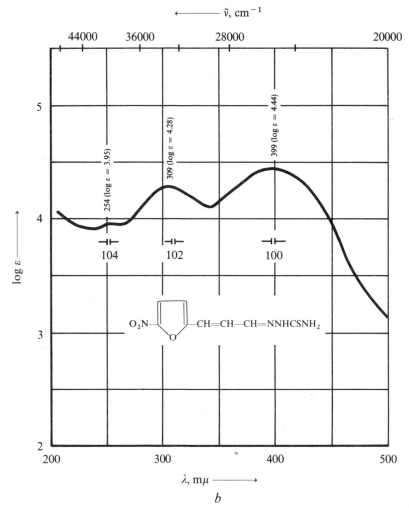

$\tilde{\nu}$, cm^{-1}

λ, mμ

b

105

Figure 42

5-(5′-Nitrofuryl-2′)-pentadien-2,4-al-1-thiosemicarbazone [62].

M. wt. 266.29; $C_{10}H_{10}N_4O_3S$; m. p. 214–219° (decomp.).

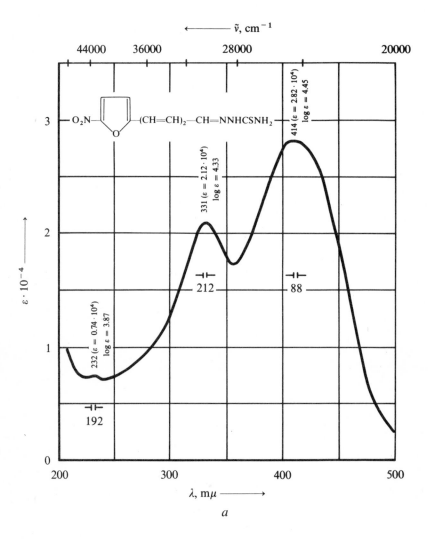

a

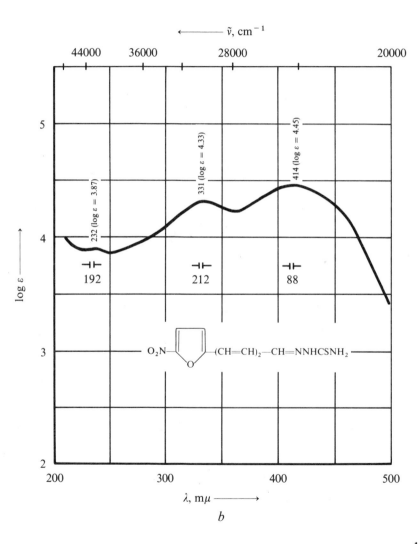

b

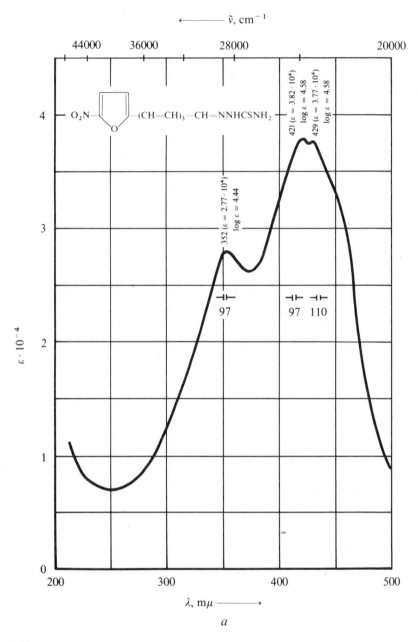

THIOSEMICARBAZONES

Figure 43

7-(5′-Nitrofuryl-2′)-heptatrien-2,4,6-al-1-thiosemicarbazone [62].
M. wt. 292.33; $C_{12}H_{12}N_4O_3S$; m. p. 223° (decomp.).

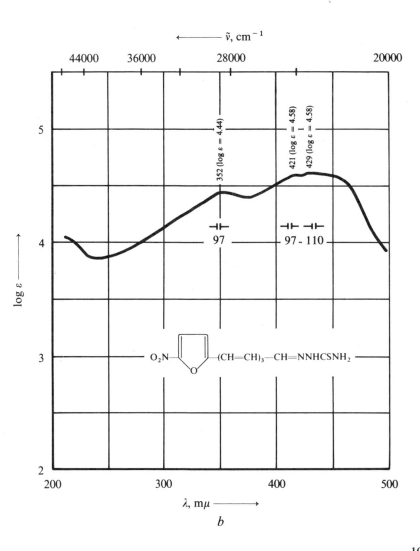

b

CYANOACETYLHYDRAZONES

Figure 44

5-Nitro-2-furfurylidene cyanoacetylhydrazone [109].
M. wt. 222.17; $C_8H_6N_4O_4$; m. p. 194–195° (decomp.).

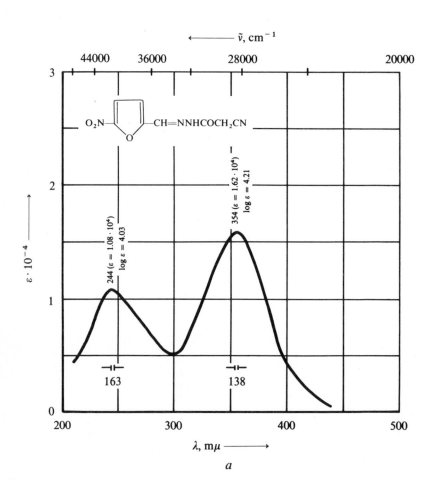

a

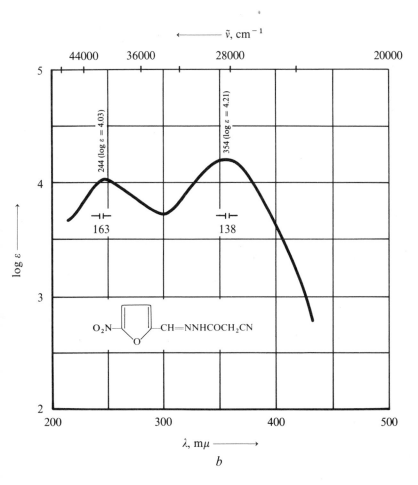

b

Figure 45

β-(5-Nitrofuryl-2)-acrylidene cyanoacetylhydrazone [108].
M. wt. 248.21; $C_{10}H_8N_4O_4$; m. p. 221° (decomp.).

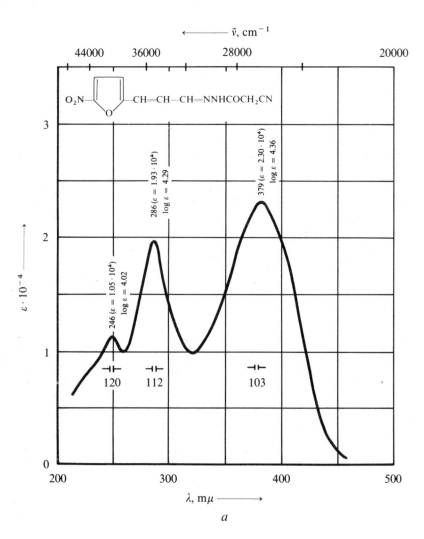

a

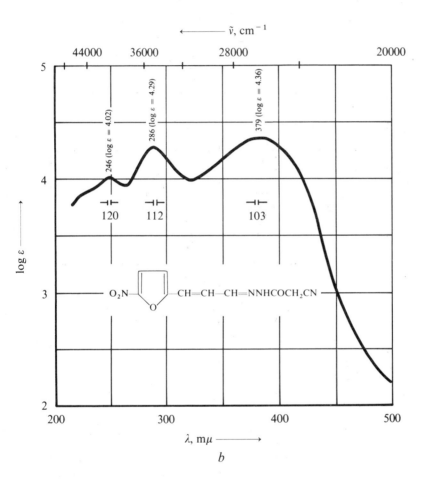

b

113

CYANOACETYLHYDRAZONES

Figure 46

5-(5′-Nitrofuryl-2′)-pentadien-2,4-al-1-cyanoacetylhydrazone [110].
M. wt. 274.24; $C_{12}H_{10}N_4O_4$; m. p. 213–214° (decomp.).

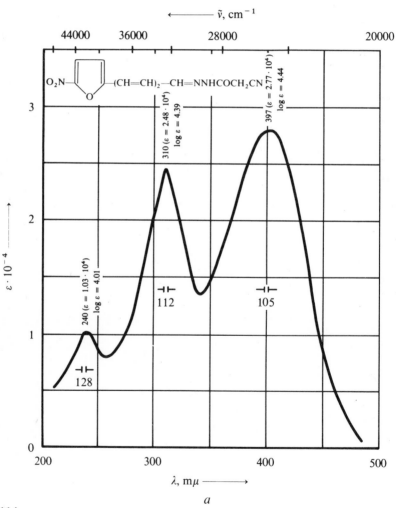

a

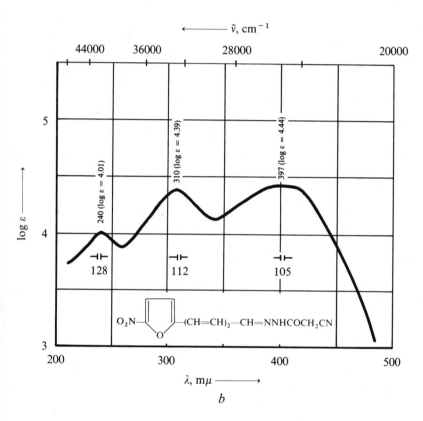

b

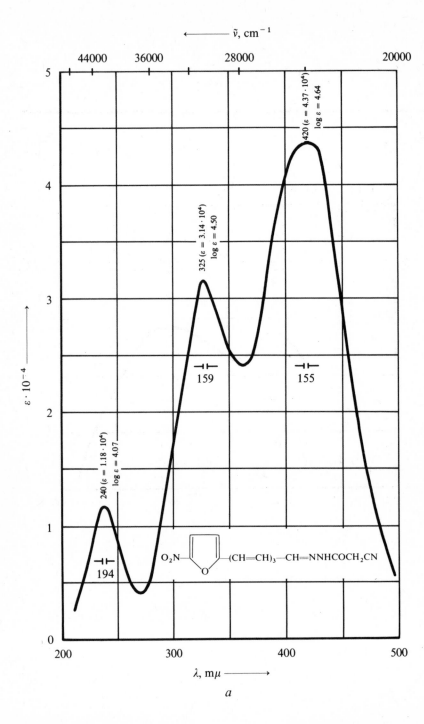

a

CYANOACETYLHYDRAZONES

Figure 47

7-(5′-Nitrofuryl-2′)-heptatrien-2,4,6-al-1-cyanoacetylhydrazone [110].
M. wt. 300.28; $C_{14}H_{12}N_4O_4$; m. p. 214° (decomp.).

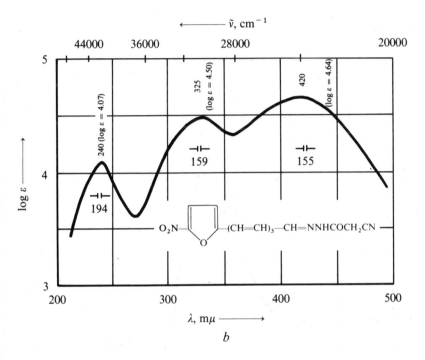

b

Figure 48

5-Nitro-2-furfurylidene isonicotinoylhydrazone [106].
M. wt. 260.22; $C_{11}H_8N_4O_4$; m. p. 252° (decomp.).

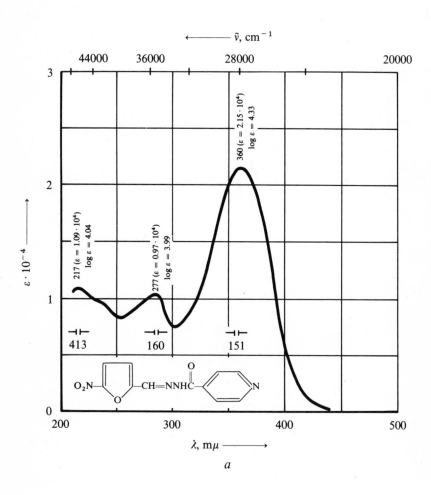

a

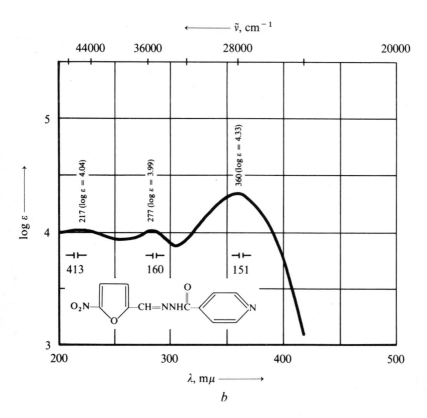

b

Figure 49

β-(5-Nitrofuryl-2)-acrylidene isonicotinoylhydrazone [108].
M. wt. 282.26; $C_{13}H_{10}N_4O_4$; m. p. 248° (decomp.).

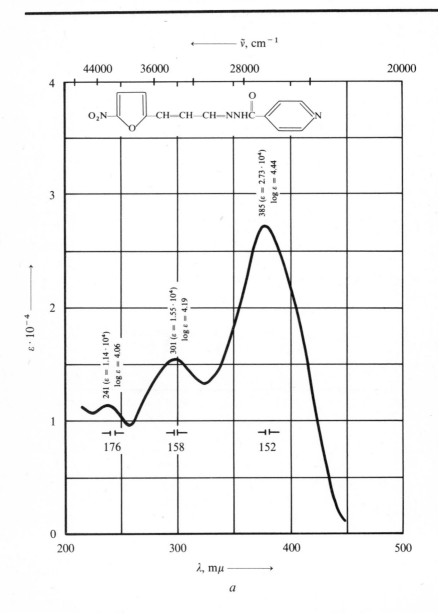

a

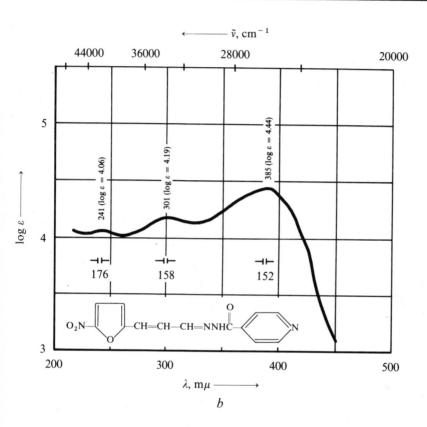

b

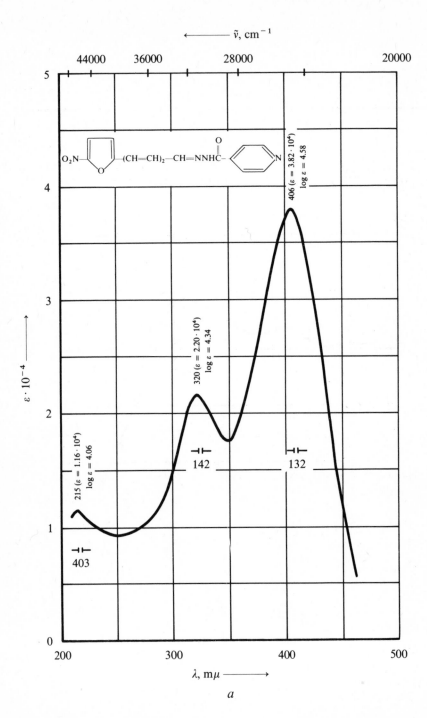

a

ISONICOTINOYLHYDRAZONES

Figure 50

5(5'-Nitrofuryl-2')-pentadien-2,4-al-1-isonicotinoylhydrazone [110].
M. wt. 312.29; $C_{15}H_{12}N_4O_4$; m. p. 237° (decomp.).

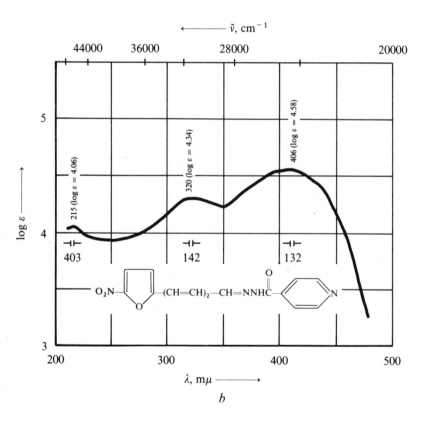

b

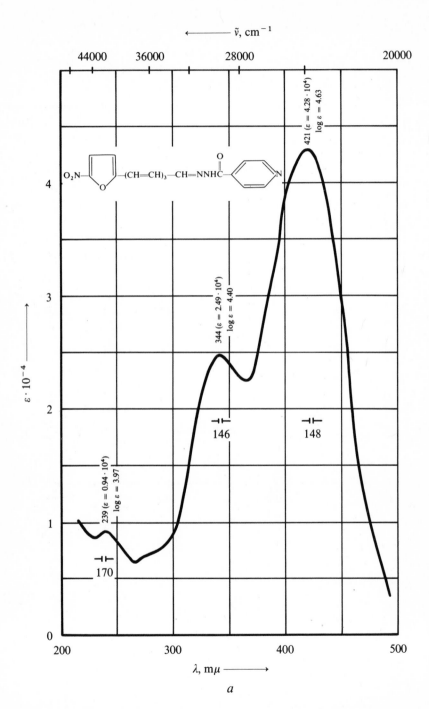

a

ISONICOTINOYLHYDRAZONES

Figure 51

7-(5′-Nitrofuryl-2′)-heptatrien-2,4,6-al-1-isonicotinoylhydrazone [110].
M. wt. 338.33; $C_{17}H_{14}N_4O_4$; m. p. 233° (decomp.).

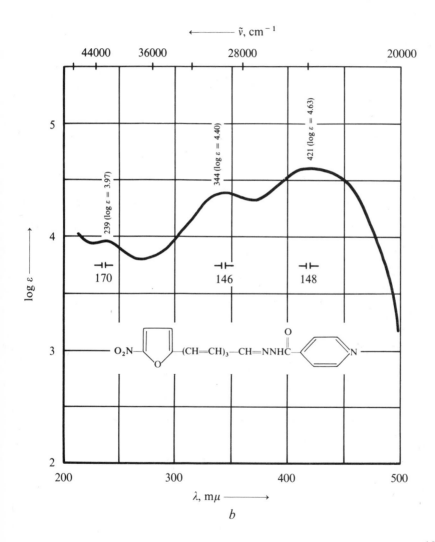

b

125

Figure 52

1-(5′-Nitro-2′-furfurylidenamino)-hydantoin [111].
M. wt. 238.17; $C_8H_6N_4O_5$; m. p. > 260° (decomp.).

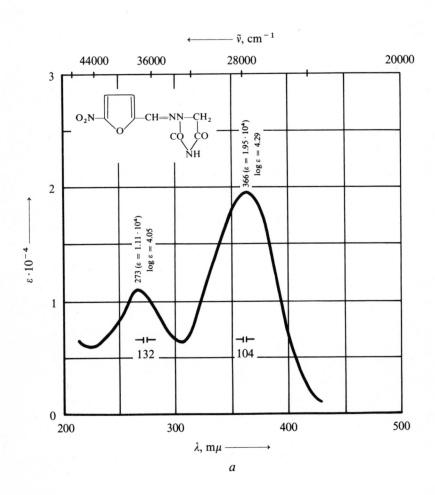

a

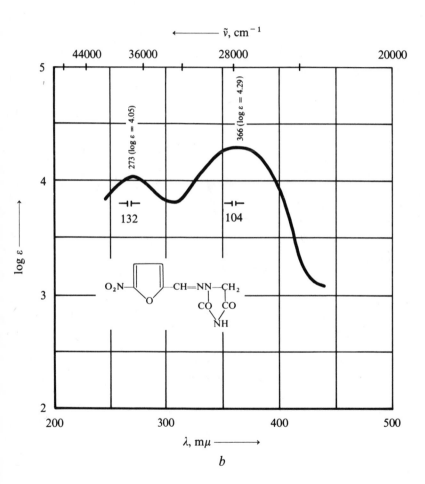

b

127

DERIVATIVES OF 1-AMINOHYDANTOIN

Figure 53

1-[β-(5'-Nitrofuryl-2')-acrylidenamino]-hydantoin [108].
M. wt. 264.21; $C_{10}H_8N_4O_5$; m. p. 260° (decomp.).

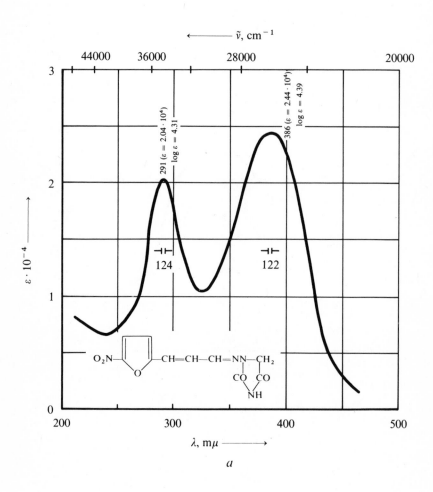

a

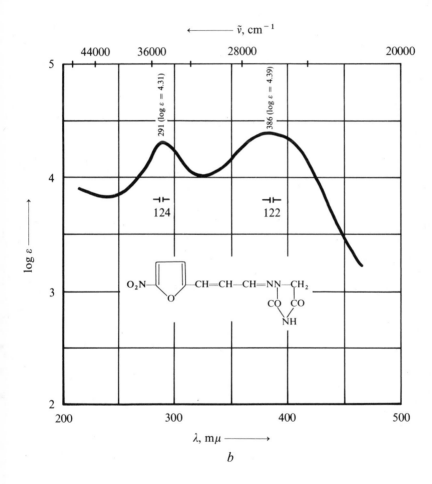

b

DERIVATIVES OF 1-AMINOHYDANTOIN

Figure 54

1-[5'-(5''-Nitrofuryl-2'')-pentadien-2',4'-al-1'-amino]-
hydantoin [110].
M. wt. 290.24; $C_{12}H_{10}N_4O_5$; m. p. $\geqslant 260°$ (decomp.).

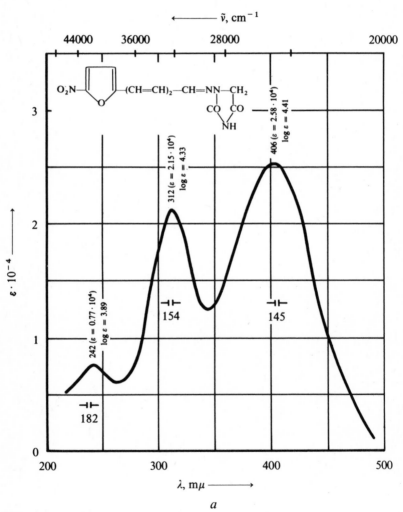

a

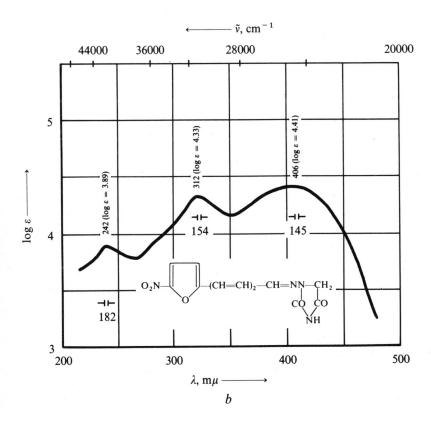

$$\tilde{v}, \text{cm}^{-1}$$

b

Figure 55

1-[7'-(5''-Nitrofuryl-2'')-heptatrien-2',4',6'-al-1'-amino]-
hydantoin [110].
M. wt. 316.28; $C_{14}H_{12}N_4O_5$; m.p. $\geqslant 285°$ (decomp.).

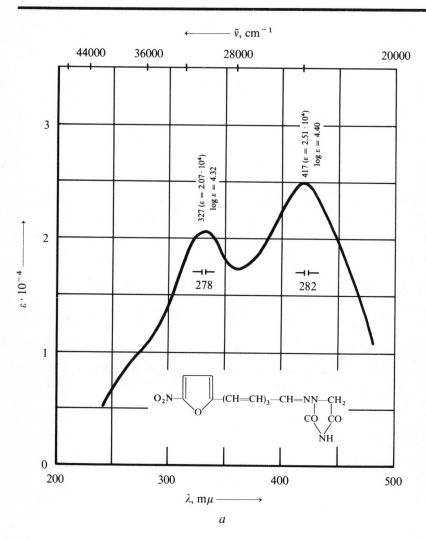

a

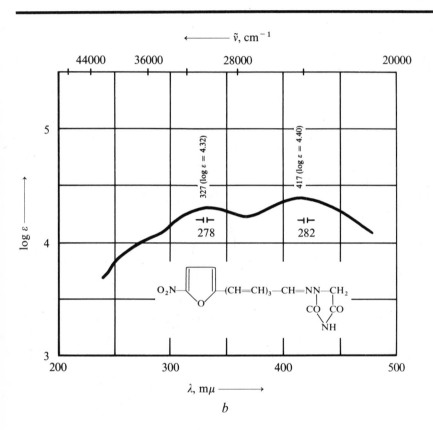

b

DERIVATIVES OF 3-AMINO-2-OXAZOLIDONE

Figure 56

3-(5'-Nitro-2'-furfurylidenamino)-2- oxazolidone [112].
M. wt. 225.17; $C_8H_7N_3O_5$; m. p. 254–256° (decomp.).

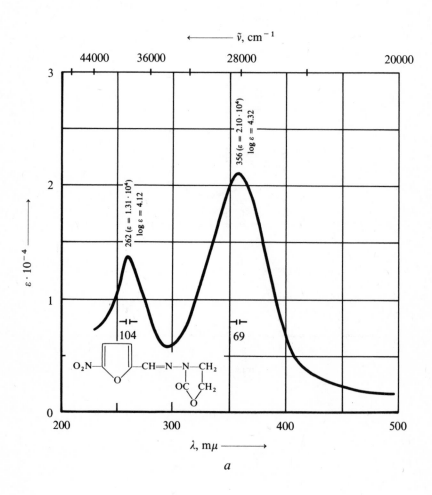

a

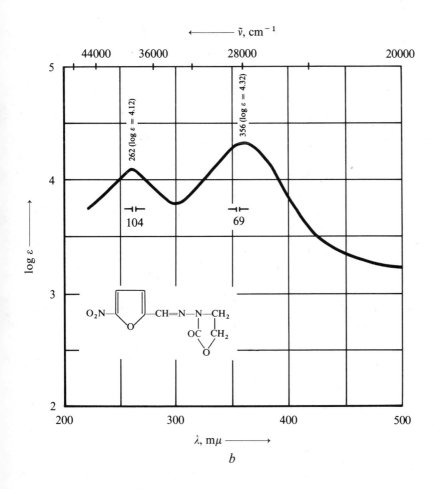

b

DERIVATIVES OF 3-AMINO-2-OXAZOLIDONE

Figure 57

3-[β-(5′-Nitrofuryl-2′)-acrylidenamino]-2-oxazolidone [108].
M. wt. 251.21; $C_{10}H_9N_3O_5$; m. p. 268° (decomp.).

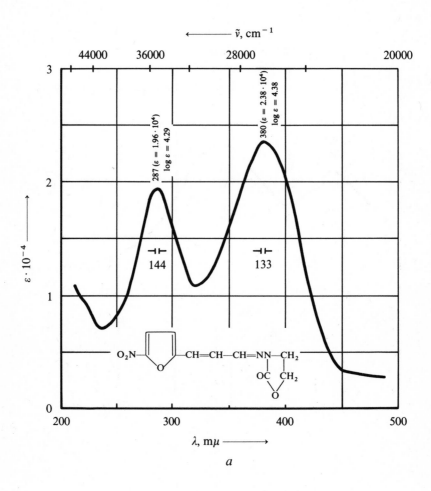

a

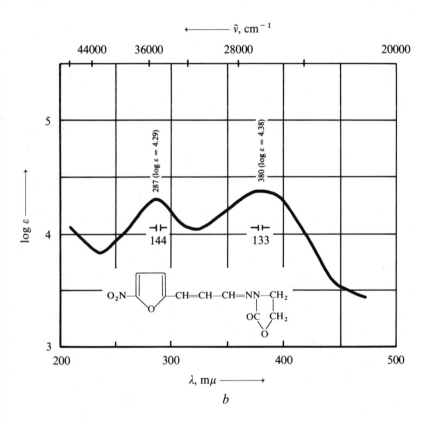

b

DERIVATIVES OF 3-AMINO-2-OXAZOLIDONE

Figure 58

3-[5′-(5″-Nitrofuryl-2″)-pentadien-2′,4′-al-1′-amino]-2-oxazolidone [110].
M. wt. 277.24; $C_{12}H_{11}N_3O_5$; m. p. 253–254° (decomp.).

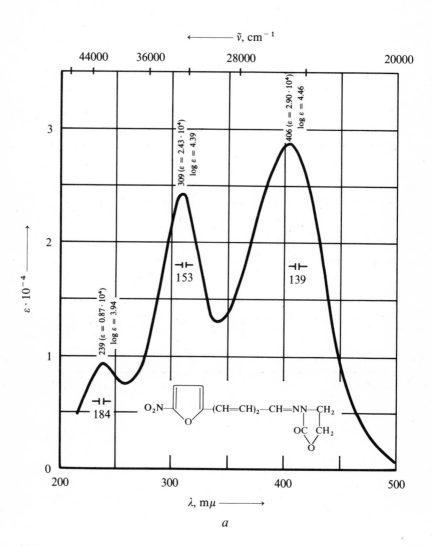

a

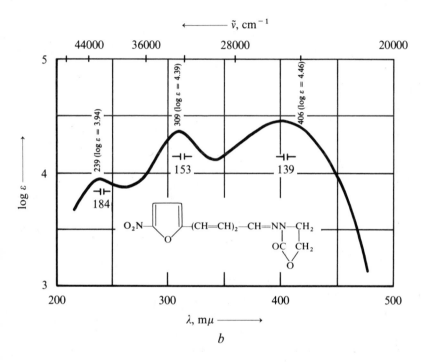

b

DERIVATIVES OF 3-AMINO-2-OXAZOLIDONE

Figure 59

3-[7'-(5''-Nitrofuryl-2'')-heptatrien-2',4',6'-al-1'-amino]-2-oxazolidone [110].
M. wt. 303.28; $C_{14}H_{13}N_3O_5$; m. p. 247° (decomp.).

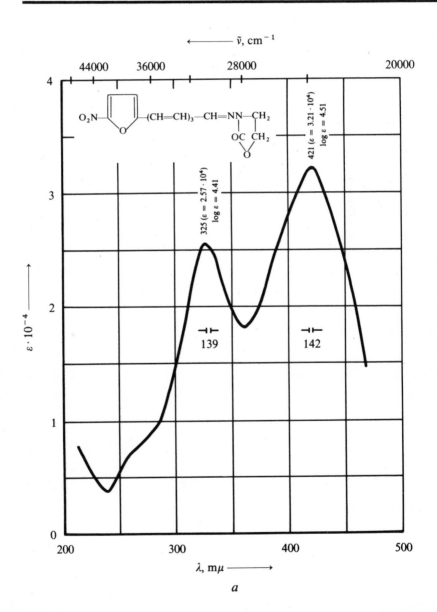

a

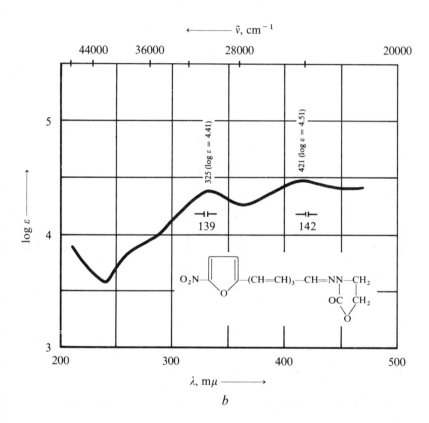

b

Figure 60

5,5′-Dinitro-2,2′-difuryl [97].

M. wt. 224.14; $C_8H_4N_2O_6$; m. p. 210° (decomp.).

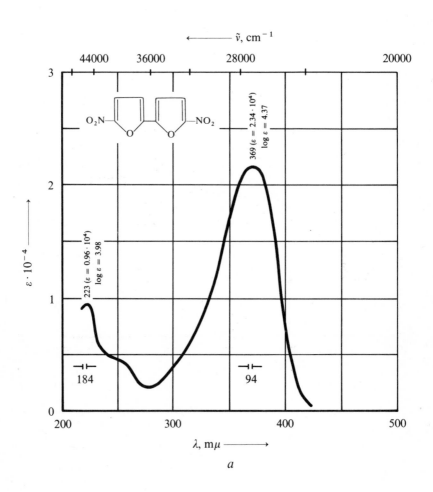

a

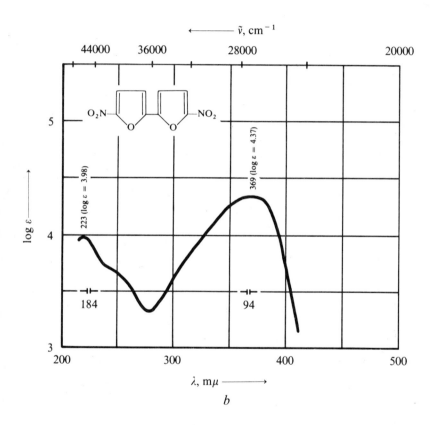

b

143

Figure 61

1,3-Bis-(5'-nitrofuryl-2')-propen-2-one-1 [49].
M. wt. 278.19; $C_{11}H_6N_2O_7$; m. p. 205–207° (decomp.).

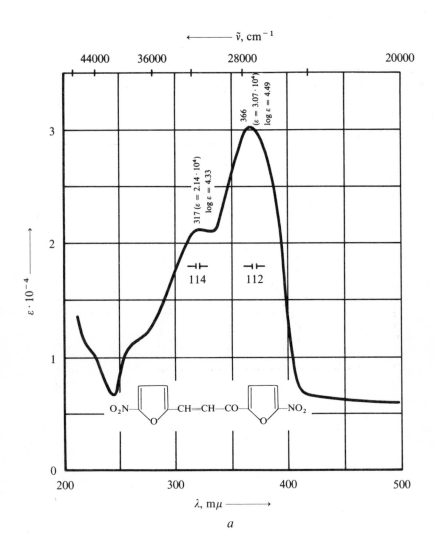

a

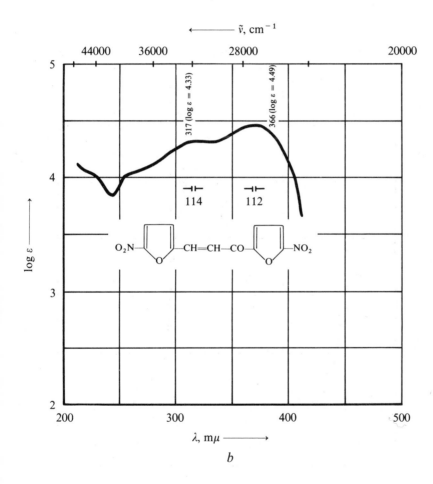

b

145

Figure 62

1,5-Bis-(5'-nitrofuryl-2')-pentadien-1,3-one-3 [49].
M. wt. 304.22; $C_{13}H_8N_2O_7$; m. p. 199° (decomp.).

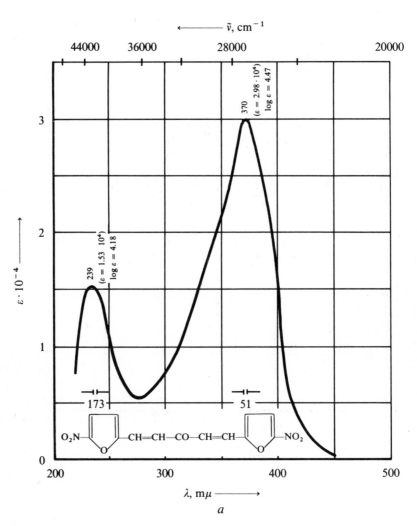

a

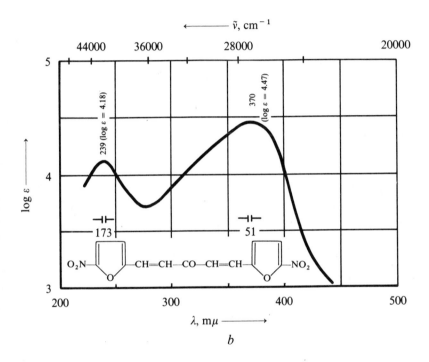

b

Bibliography

1. SCHEELE, K.W. *Saemtliche physische und chemische Werke*, **2**, 265.—Berlin. 1793; *Beilstein*, No. 18. 272.
2. DÖBEREINER, I.W.—*Ann.*, **3**, 141. 1832.
3. LIMPRICHT, H.—*Ber.*, **3**, 90. 1870.
4. KLINKHARDT, A.—*J. prakt. Chem.*, **25**, 41. 1882.
5. DODD, M.C., W.B. STILLMAN, M. ROYS, and C. CROSBY.—*J. Pharmacol. and Exptl. Therap.*, **82**, 11. 1944.
6. HILLER, S.A. and J.A. EIDUS.—*Izvestiya AN Latv. SSR*, No. 8, 1223. 1951.
7. HILLER, S.A., J.A. EIDUS, and N.O. SALDABOL.—*Izvestiya AN SSSR*, **17**, 708. 1953; *Izvestiya AN SSSR*, **19**, No. 5, inside back cover. 1955.
8. EIDUS, J.A. and L. MUTSENIETSE.—*Izvestiya AN Latv. SSR*, No. 11, 65. 1961.
9. EIDUS, J.A., K.K. VENTERS, and S.A. HILLER.—*Doklady AN SSSR*, **141**, 655. 1961.
10. EIDUS, J.A., K.K. VENTERS, and S.A. HILLER.—In: *Fizicheskie problemy spektroskopii*, **1**, 276. Moscow, Izdatel'stvo AN SSSR. 1962.
11. EIDUS, J.A. and I.V. ZUIKA.—*Izvestiya AN Latv. SSR*, phys., techn. sci. ser., No. 2, 75. 1965.
12. EIDUS, J.A., T.N. POLKO, and YU.K. YUR'EV.—*Izvestiya AN Latv. SSR*, No. 2, 63. 1963.
13. BOBOVICH, YA.S. and J.A. EIDUS.—*Optika i Spektroskopiya*, **16**, 424. 1964.
14. EIDUS, J.A., K.K. VENTERS, and I.V. ZUIKA.—*KhGS*, 402. 1967.
15. HILLER, S.A., J.A. EIDUS, and N.O. SALDABOL.—*Seventh All-Union Conference on Spectroscopy*. Abstracts of papers. Moscow. 1952.
16. EIDUS, J.A., K.K. VENTERS, and S.A. HILLER.—*Eighth All-Union Conference on Spectroscopy*. Abstracts of papers. Leningrad. 1960.
17. EIDUS, J.A., K.K. VENTERS, and S.A. HILLER.—*Inter-University Conference on the Theory of Chemical Structure, Kinetics and Reactivity*. Abstracts of papers. Riga. 1961.

18. Eidus, J.A., S.A. Hiller, and K.K. Venters.—*Nauchno-metodicheskaya konferentsiya Latviiskogo gosudarstvennogo universiteta im. P. Stuchki.* Proceedings of the Conference. Riga. 1961.

19. Eidus, J.A., K.K. Venters, and S.A. Hiller.—*Second All-Union Scientific Conference on the Chemistry of Furan Compounds.* Abstracts of papers. Saratov. 1952.

20. Eidus, J.A.—*Third Conference on Spectroscopy of the Lithuanian SSR.* Abstracts of papers. Vilna. 1964.

21. Eidus, J.A.—*All-Union Symposium on the Intensity and Shape of Spectral Lines.* Abstracts of papers. Krasnoyarsk. 1964.

22. Matsen, F.—In: *Applications of Spectroscopy in Chemistry.* [Russian translation. 1959.]

23. Thompson, H. and R.B. Temple.—*Trans. Faraday Soc.,* **41**, 27. 1945.

24. Bak, B. and D. Christensen.—*J. Mol. Spectr.,* **9**, 124. 1962.

25. Schomaker, V. and L. Pauling.—*J. Amer. Chem. Soc.,* **61**, 1769. 1939.

26. Dunlop, A. and L. Peter. *The Furans.*—New York, Reinhold Publ. 1953.

27. Beach, J.—*J. Chem. Phys.,* **9**, 54. 1940; *C.A.,* **35**, 1279. 1940.

28. Landolt-Börnstein. *Physikalisch-Chemische Tabellen,* **3**, 297. Berlin. Springer-Verlag. 1951.

29. Gillam, A. and E. Stern. *An Introduction to Electronic Absorption Spectroscopy in Organic Chemistry.*—London, Edward Arnold Publ. 1954.

30. Menczel, S.—*Z. phys. Chem.,* **125**, 161. 1927.

31. Pickett, L.W.—*J. Chem. Phys.,* **8**, 293. 1940.

32. Eidus, J.A. *Stroenie i spektry 5-nitrofuranov (Structure and Spectra of 5-Nitrofurans).* Candidate's Dissertation. Riga. 1964.

33. Horváth, G. and A.I. Kiss.—*Spectrochim. acta,* **23A**, 921. 1967.

34. Padleye, M.R. and S.R. Desai.—*Proc. Phys. Soc.,* **65**, 298. 1952.

35. Kasha, M.—*Disc. Faraday Soc.,* **9**, 14. 1950.

36. Pickett, L.W., M. Muntz, and E.M. McPherson.—*J. Amer. Chem. Soc.,* **73**, 4862. 1941.

37. Feofilov, P.P. *Polyarizovannaya lyuminestsentsiya atomov, molecul i kristallov (Polarized Luminescence of Atoms, Molecules and Crystals).* Moscow, Fizmatgiz. 1959.

38. Boryniec, A. and L. Marchlevski.—*Bull. internat. Acad. Pol. Sci., Lettres,* **A**, 392. 1931; *C.A.,* **27**, 229. 1933.

39. Abe, S.—*J. Chem. Soc. Japan, Industr. Chem. Sec.,* **55**, 901. 1934; *C.A.,* **29**, 305. 1935.

40. Hughes, E.C. and J.R. Johnson.—*J. Amer. Chem. Soc.,* **53**, 737. 1931.

41. Raffauf, R.F.—*J. Amer. Chem. Soc.,* **72**, 753. 1950.

42. Kortuem, G.—*Z. phys. Chem.,* **B43**, 271. 1939.

43. Kortuem, G.—*Z. Elektrochem.,* **47**, 55. 1941.

44. MASAKI, K.—*Bull. Chem. Soc. Japan*, **11**, 712. 1936; *C.A.*, **31**, 1700. 1937.
45. MURELL, J.N. *The Theory of the Electronic Spectra of Organic Molecules*, pp. 158, 184.—London, Methuen. 1963.
    MULLIKEN, R.S.—*J. Chem. Phys.*, **3**, 564. 1935.
46. BRAUDE, E.A., E.R.H. JONES, and G.G. ROSE.—*J. Chem. Soc.*, 1104, 1947.
47. EDWARDS, W.R. and C.W. TATE.—*Analyt. Chem.*, **23**, 826. 1951.
48. VENTERS, K.K., J.A. EIDUS, D.D. LOLYA, and S.A. HILLER.—*Conference on the Chemistry of Oxygen-Containing Heterocycles*. Odessa. 1966.
49. VENTERS, K.K., J.A. EIDUS, S.A. HILLER, and D.O. LOLYA.—*KhGS*, 405. 1968.
50. WYNBERG, H. and J.W. VAN REIJENDAM.—*Reçueil trav. chim.*, **86**, 381. 1967.
51. DOBRINSKAYA, A.A., M.B. NEIMAN, L.N. POLKANOVA, and R.B. PROTSENKO.—*Doklady AN SSSR*, **63**, 543. 1948.
52. DOBRINSKAYA, A.A. and M.B. NEIMAN.—*Izvestiya AN SSSR*, phys. ser., **14**, 529. 1950.
53. KUHN, H.—*Helv. chim. acta*, **31**, 1441. 1948.
54. HAUSSER, K.W., R. KUHN, A. SMAKULA, and M. HOFFER.—*Z. phys. Chem.*, **B29**, 371. 1935.
55. BLOUT, E.R. and M. FIELDS.—*J. Amer. Chem. Soc.*, **70**, 189. 1948.
56. STRADIŅŠ, J.P., S.A. HILLER, and YU.K. YUR'EV.—*Doklady AN SSSR*, **129**, 816. 1959.
57. HAUSSER, K.W., R. KUHN, A. SMAKULA, and A. DEUTSCH.—*Z. phys. Chem.*, **B29**, 378. 1935.
58. PAPPALARDO, G.—*Gazz. chim. ital.*, **89**, 551. 1959.
59. BOLOVICH, YA.S. and V.V. PEREKALIN.—*Doklady AN SSSR*, **121**, 1028. 1958.
60. CLAUSON–KAAS, N. *Reaction of the Furan Nucleus; 2,5-Dialkoxy-2,5-dihydrofurans and 2,5-Diacetoxy-2,5-dihydrofuran.*—Copenhagen. 1947.
61. VENTERS, K.K. and S.A. HILLER.—*Nitro Compounds (Proceedings of the International Symposium held at the Polish Academy of Sciences, Warsaw, 18–20 Sept. 1963)*, p. 71. Warsaw, Wydawnictwa Naukowo technicze. 1964.
62. SAIKACHI, H. and H. OGAWA.—*J. Amer. Chem. Soc.*, **80**, 3642. 1958.
63. HAUSSER, K.W.—*Z. techn. Phys.*, **1**, 10. 1934.
64. HAUSSER, K.W., R. KUHN, and K.H. KREUCHEN.—*Z. phys. Chem.*, **B29**, 363. 1935.
65. HAUSSER, K.W., R. KUHN, and G. SEITZ.—*Z. phys. Chem.*, **B29**, 391. 1935.
66. HAUSSER, K.W., R. KUHN, and A. SMAKULA.—*Z. phys. Chem.*, **B29**, 384. 1935.
67. LEWIS, G.N. and M. CALVIN.—*Chem. Revs.*, **25**, 273. 1939.
68. KUHN, R.—*J. Chem. Soc.*, 605. 1938.
69. KUHN, H.—*J. Chem. Phys.*, **17**, 1198. 1949.
70. PLATT, J.R.—*J. Chem. Phys.*, **25**, 80. 1956.

71. BAYLISS, N.S.—*J. Chem. Phys.*, **16**, 840. 1948.
72. SIMPSON, W.T.—*J. Chem. Phys.*, **16**, 1124. 1948.
73. SOMMERFELD, A. and H. BETHE. *Handbuch der Physik*, 2nd ed., **24**, part 2.— Berlin, Springer-Verlag. 1933.
74. SCHMIDT, O.—*Z. phys. Chem.*, **B47**, 1. 1940.
75. KUHN, H.—*Z. Elektrochem.*, **53**, 165. 1949.
76. KUHN, H.—*Helv. chim. acta*, **32**, 2247. 1949.
77. KUHN, H.—*Helv. chim. acta*, **34**, 1308. 1951.
78. HUBER, W., J.F. HORNIG, and H. KUHN.—*J. Chem. Phys.*, **25**, 1296. 1956.
79. HUBER, W., J.F. HORNIG, and H. KUHN.—*Z. phys. Chem.*, **NF9**, 1. 1956.
80. KUHN, H.—*Angew. Chem.*, **71**, 93. 1959.
81. BAER, F., W. HUBER, G. HANDSHIG, and H. KUHN.—*J. Chem. Phys.*, **32**, 470. 1960.
82. KUHN, H., W. HUBER, G. HANDSHIG, H. MARTIN, F. SCHAERFER, and F. BAER.—*J. Chem. Phys.*, **32**, 467. 1960.
83. MORSE, P.M. and H. FESHBACH. *Methods of Theoretical Physics.*—New York, McGraw-Hill. 1953.
84. LANDAU, L. and E. LIFSHITS.—*Kvantovaya mekhanika (Quantum Mechanics).* **1**, pp. 89–90. Moscow–Leningrad, GITTL. 1948.
85. MULLIKEN, R.S.—*J. Chem. Phys.*, **7**, 121. 1939.
86. HERZBERG, K.F.—*J. Chem. Phys.*, **10**, 508. 1942.
87. HERZBERG, K. and A.L. SKLAR.—*Revs. Mod. Phys.*, **14**, 294. 1942.
88. LENNARD–JONES, J.E.—*Proc. Roy. Soc.*, **A158**, 280. 1937.
89. ROBERTS, J.D. *Notes on Molecular Orbital Calculations.*—New York. 1962.
90. SAPPENFIELD, D.S. and M. KREEVOY.—*Tetrahedron*, **19**, Suppl. 2., 157. 1963.
91. WEISSBERGER, A. editor, *Technique of Organic Chemistry*, **7**, *Organic Solvents.*—N.Y. Interscience. 1955.
92. MARQUIS, R.—*Ann. chim. phys.*, (8) **4**, 196. 1905.
93. RINKES, I.J.—*Reçueil trav. chim.*, **49**, 1118. 1930.
94. GILMAN, H. and G.F. WRIGHT.—*J. Amer. Chem. Soc.*, **52**, 2550. 1930.
95. HILLER, S.A. and K.K. VENTERS.—*Izvestiya AN Latv. SSR*, No. 12, 115. 1959.
96. VENTERS, K.K., S.A. HILLER, and A.A. LAZKYN'SH.—*Izvestiya AN Latv. SSR*, No. 5, 87. 1961.
97. SASAKI, T.—*Bull. Inst. Chem. Res., Kyoto Univ.*, **33**, 39. 1955.
98. SALDABOL, N.O. and S.A. HILLER.—*Izvestiya AN Latv. SSR*, No. 10, 101. 1958.
99. VENTERS, K.K., S.A. HILLER, and N.O. SALDALBOL.—*Izvestiya AN Latv. SSR*, No. 8, 99. 1959.
100. VENTERS, K.K., S.A. HILLER, V.F. KUCHEROV, V.V. TSIRULE, and A.M. KARTLINYA.—*Doklady AN SSSR*, **140**, 1073. 1961.

101. KURGAN, B.V., A.A. GRUZE, and S.A. HILLER.—In: *Metody polucheniya khimicheskikh reaktovov i preparatov,* **13**, 80. Moscow, IREA. 1965.

102. VENTERS, K.K., S.A. HILLER, and V.V. TSIRULE.—*Izvestiya AN Latv. SSR,* chem. ser., 131. 1962.

103. MARQUIS, R.—*Compt. rend.,* **137**, 520. 1903.

104. NISHIDA, S., T. SATO, and Y. SATO.—*Repts. Scient. Res. Inst.,* **31**, 430. 1955.

105. PONOMAREV, A.A. *Sintezy i reaktsii furanovykh veshchestv (Syntheses and Reactions of Furan Preparations),* 81.—Saratov, Izdatel'stvo Saratovskogo Universiteta, 1960.

106. HILLER, S.A.—In: *Furatsilin i opyt ego primeneniya,* p. 7. Riga, Izdatel'stvo Latv. SSR, 1953.

107. TAKAHASHI, T., H. SAIKACHI, SH. YOSHINA, and CH. MIZUNO.—*J. Pharmac. Soc. Japan,* **69**, 284. 1949; *C.A.,* **44**, 5372. 1950.

108. HILLER, S.A.—In: *Voprosy ispol'zovaniya pentozansoderzhashchego syr'ya,* p. 451. Riga, Izdatel'stvo AN Latv. SSR, 1958.

109. DELMAR, G.S. and E.N. MACALLUM. Canadian Patent No. 594397. 1960; *C.A.,* **54**, 22681. 1960.

110. HILLER, S.A., S.P. ZAEVA, K.K. VENTERS, L.N. ALEKSEEVA, L.V. KRUZMETRA, and S.K. GERMANE.—*KhGS,* 187. 1965.

111. HAYES, K.J.—U.S. Patent No. 2610181. 1952; *C.A.,* **47**, 6980. 1953.

112. HILLER, S.A. and R.YU. KALNBERG.—In: *Furazolidon.* Riga, Izdatel'stvo AN Latv. SSR, 1962, 5.

Printed in Israel
Manufactured at the Israel Program for Scientific Translations, Jerusalem

RETURN TO ➡

CHEMISTRY LIBRARY
100 Hildebrand Hall • 642-3753

LOAN PERIOD 1	2	3
		1 MONTH
4	5	6

ALL BOOKS MAY BE RECALLED AFTER 7 DAYS
Renewable by telephone

DUE AS STAMPED BELOW

FORM NO. DD5

UNIVERSITY OF CALIFORNIA, BERKELEY
BERKELEY, CA 94720-6000